MODELLING THE EFFICIENCY OF FAMILY
AND HIRED LABOUR

To my parents

Modelling the Efficiency of Family and Hired Labour

Illustrations from Nepalese Agriculture

PREM JUNG THAPA
Economics Program
Research School of Social Sciences
Australian National University

and

Institute for Integrated Development Studies
Kathmandu

ASHGATE

Published by
Ashgate Publishing Limited
Gower House
Croft Road
Aldershot
Hampshire GU11 3HR
England

Ashgate Publishing Company
Suite 420
101 Cherry Street
Burlington, VT 05401-4405
USA

Ashgate website: http://www.ashgate.com

British Library Cataloguing in Publication Data
Thapa, Prem Jung
    Modelling the efficiency of family and hired labour
    1.Agricultural labourers - Nepal 2.Agriculture - Economic
    aspects - Nepal 3.Family farms - Nepal 4.Division of labour
    - Nepal
    I.Title
    338.1'095496

Library of Congress Cataloging-in-Publication Data
Thapa, Prem Jung, 1954-
    Modelling the efficiency of family and hired labour : illustrations from Nepalese
    agriculture / Prem Jung Thapa.
        p. cm.
    Includes bibliographical references and index.
    ISBN 0-7546-1853-6 (alk. paper)
    1. Agricultural labourers--Statistical methods. 2. Family farms--Statistical methods. 3.
    Agricultural labourers--Nepal--Statistical methods. 4. Family farms--Nepal--Statistical
    methods. I. Title.

HD1521 .T465 2002
338.1'6'095496--dc21   2002028121

ISBN 0 7546 1853 6

Printed in Great Britain by Antony Rowe Ltd, Chippenham, Wiltshire

Contents

List of Figures

List of Tables

Preface

Is hired labour a perfect substitute for family labour as an input in farm production? Are these two types of labour equally productive? A negative answer to either question has important implications for the analytical modelling and empirical estimation of the labour demand and supply decisions of farm households, and for deriving welfare measures based on the implicit valuation (shadow price) of family labour. Significant policy implications also arise with imperfect substitutability and/or productivity differences between family and hired labour because aggregate outcomes in the agricultural sector will be sensitive to changes in the distribution of individual endowments of land and labour.

This study uses a production function based approach to test for heterogeneity between family and hired labour as inputs in crop production in the *tarai* (southern lowland) region of Nepal. It develops an appropriate analytical framework and carries out the empirical estimation of an aggregate farm household model that allows for heterogeneity between family and hired labour. A sequential estimation strategy is adopted. The labour heterogeneity detected in the first step of the production function estimation is incorporated, at the second step, in the labour supply estimation in a theoretically consistent manner. The methodological novelty is to relate the shadow wage rate for family labour to the observed market wage rate for hired labour, adjusted for the differential productivity of family and hired labour detected in the production function estimates.

The estimation strategy adopted in this book also provides a useful extension of the literature on farm household models. The conventional approach to estimating a farm household model treats family and hired labour *a priori* as homogenous inputs in order to preserve the recursive (separable) property of the model that allows the production and consumption decisions of the farm household to be estimated separately. The analytical framework presented here illustrates the conditions under which the recursive structure of the conventional farm household model can be preserved even when family and hired labour are allowed to be heterogeneous inputs. The recursive property is affected not just by the nature of the labour heterogeneity but also by the market exposure of the farm household in the hired labour market. Whether a farm household is a net seller or buyer of labour in the local labour market becomes a significant determinant of the effective wage rate it faces for its family labour supply.

The final empirical results presented for farm households of the *tarai* region of Nepal show that allowing for possible labour heterogeneity does make a difference in the estimates of the factor demand and labour supply elasticities that define a farm household model. The estimates of these elasticities with labour heterogeneity are not only within reasonable bounds and have expected signs, but in several instances they also differ significantly from the estimates based on the conventional approach that assumes labour homogeneity.

The evidence presented in this book for an efficiency difference between family and hired labour, with family labour being more productive, also sheds light on a celebrated stylized fact of agriculture in developing countries: the inverse relationship between farm size and intensity of labour use, which often leads to an inverse relationship between farm size and yield. The common traditional explanation for the inverse relationship relies on rural labour market imperfections which lead smaller sized farms to apply too much family labour to their own farms, relative to the prevailing market wage rate for agricultural labour. A higher efficiency of family labour provides an alternative explanation for the lower intensity of labour application on the bigger farms (that will be more reliant on hired labour), without having to resort to explanations based on the existence of labour or other factor market failures.

This book does not investigate directly the underlying causes or sources of the efficiency difference between family and hired labour detected in the production function estimation results. It would be useful hence to direct further research to discriminate between alternative explanations for the underlying source of the labour heterogeneity so that firmer implications of labour heterogeneity for land reform and rural labour market policies can be developed.

Acknowledgements

I gratefully acknowledge the support and encouragement received from many individuals and institutions, without which it would have not been possible to complete this book. The material in this book was initially prepared for a Ph.D. thesis at the Research School of Pacific and Asian Studies of the Australian National University (ANU). I owe a very special debt of gratitude to Peter Warr, my supervisor, for the academic and personal support I received over the period of conceptualizing and completing this research. I also received invaluable advice and help from the other two members of my supervision committee, George Fane and K. Kalirajan. I wish to acknowledge their contribution in nurturing my research and in improving the analysis and presentation through their critical comments and suggestions.

Others who have read parts of these materials at different stages and provided helpful comments are Prema-chandra Athukorala, Xin Meng, Hal Hill and Chris Manning. I have also benefited from consultations and advice received on specific aspects of this research from Adrian Pagan, Dennis Hare and Dale Squires. I thank all of the above; but all remaining errors and omissions are my responsibility. Several parts of the material of this book were also presented in the Third Labour Econometrics Workshop organized by the Centre for Economic Policy Research of the ANU and the University of Tasmania in April 2000 at Hobart, and I benefited from the comments received, particularly from Tom Crossley.

I also wish to record the considerable support and encouragement I received from the Institute for Integrated Development Studies (IIDS), Kathmandu, from where I was on study leave during the period of my ANU enrolment. I express my gratitude especially to Mr. Kul Shekhar Sharma, Dr. Meena Acharya and Dr. Mohan Man Sainju for facilitating my ANU enrolment and the completion of this research.

I am very grateful to Nepal Rastra Bank for giving me access to and help in processing the household level data files from the Multipurpose Household Budget Survey of Nepal. A very special thanks goes to Mr. Bhaskar Risal, who I have bothered for help with the survey data and codebooks even years after his official position as Project Director of the survey had elapsed. I would not have been able to perform the estimations reported in this book without his unrelenting support.

Some sections of the Rastra Bank survey data were also processed at IIDS from the original survey questionnaires. I am grateful again to IIDS for organizing and financially supporting this additional data processing. Many individuals contributed to this work. I wish to express my gratitude in particular to Dr. Krishna Dev Bhatta, Asosthama Pokhrel, Shanker Aryal, Gajendra Shrestha and Ram Kumar Biswokarma.

I have received financial support from many sources during the period of my Ph.D. studies at the Australian National University and the subsequent period I have spent in preparing this book. I gratefully acknowledge the financial support provided by the World Bank Graduate Scholarship Program, the Australian National University, IIDS, and by the Ford Foundation, New Delhi for funding various components of my ANU enrolment. I am also grateful for the assistance and research time made available to me by the Centre for Regional Economics (CREA) of the University of Tasmania where most of the revisions for this book was completed. I thank John Madden, the then Director of CREA, for his personal support; and also Lydia Parish and Robyne Kerr, both at the University of Tasmania, for their able assistance in the preparation of the manuscript and map pages. I also gratefully acknowledge the excellent facilities and support system for research I have received in the Economics Program of the ANU's Research School of Social Sciences where this book was finally completed.

I am also grateful for the support I have received from the editorial and production staff at Ashgate, especially from Cilla Kennedy, Carolyn Court and Pauline Beavers.

My deep gratitude goes to my wife, Bijaya, for her unflagging support and sustenance, and the admirable patience that she has shown during the long period over which this study has been completed. I also gratefully acknowledge the cheerful encouragement and appreciation of my research I have received from my uncle, C. S. Thapa, from which I drew much of the inspiration needed to complete my degree and this book. Finally, a special word of thanks to Mac Brown, Leo Rose and Pranab Bardhan for getting me started on this journey.

Prem J. Thapa

List of Abbreviations

AES	Allen partial elasticity of substitution
AIDS	Almost Ideal Demand system
CBS	Central Bureau of Statistics, Kathmandu, Nepal
CES	constant elasticity of substitution production function
Cons.	consumption
CRTS	constant returns to scale
CRINT	cropping intensity
CSS	complete strong separability of factor inputs
DES	direct elasticity of substitution between two factors
DF, df	degrees of freedom
Eq.	equation
FIML	full information maximum likelihood (estimator)
GL	generalized linear production function (Diewert)
HEC	Hicksian elasticity of complementarity
HES	Hicksian elasticity of substitution (the reciprocal of HEC)
HMG/N	His Majesty's Government of Nepal
ICRISAT	International Crop Research Institute for the Semi-Arid Tropics
IIDS	Institute for Integrated Development Studies, Kathmandu
IMR	inverse Mill's ratio
MPHBS	Multipurpose Household Budget Survey (conducted by Nepal Rastra Bank)
MRS	marginal rate of substitution (utility function elements)
MRTS	marginal rate of technical substitution (production inputs)
NL	non-linear least squares
NRB	Nepal Rastra Bank
OLS	ordinary least squares
PSS	partial strong separability of factor inputs
Rs.	Rupees (Nepalese currency unit)
TL	translog production function
VES	variable elasticity of substitution production function
WS	weak separability of factor inputs in production function

Guide to Notation

A	land input (generic usage)
A_c	net farm cultivated area
Ae	effective land input
A_g	gross farm harvested area
B	bullock power input
C	composite consumption good in the household utility function
E	non-labour income of farm household, excluding farm profit
E*	non-labour income of farm household, including farm profit
F	family labour workdays allocated to own farm cultivation
H	hired labour workdays allocated to farm cultivation
$\mathcal{L}$	leisure in the household utility function
L	labour demand or labour supply (based on the context)
Ld, L_d	labour demand
Le	effective labour input in farm cultivation
Ls, L_s	labour supply
M	*in Chapter 6*: material inputs in crop production
M	*elsewhere*: family labour days worked in the off farm labour market for a wage
NLY	household level non-labour income
p	price of farm output
π	farm profit: pure return to farm household from land ownership
p_c	price of consumption good
PNLY	per capita non-labour income
Q	aggregate farm output quantity
θ	ratio of marginal productivity of hired labour to family labour
$\theta*$	ratio of the marginal product of hired labour to family labour at the optimum labour allocation in farm cultivation
T	total time endowment
w	wage rate received by family labour for off farm labour supply
w^h	wage rate paid to hired labour to work on the family farm
w*	effective wage rate applicable for on farm family labour supply
wr	ratio of the female hired labour wage rate to the male wage rate

1 Introduction

1.1 Motivation

Family based agricultural/farm households are the principal form of economic organization in most developing countries. Their economic decisions on farm production, labour use and consumption and leisure choices can be quite complex because the farm/household acts both as a producer and consumer unit. In conventional economic analysis production and consumption decisions are treated independently. A firm's decisions on optimal input use are not affected by the utility maximizing consumption decisions of the owners of the firm. But in the traditional agriculture setting, where the farm family supplies most of the production inputs, household preferences over the consumption bundle, including leisure, can directly affect production decisions. Conversely, production technology and outcomes can directly affect consumption choices, in addition to the conventional income effects.

When decisions on production and consumption are systematically inter-linked, behavioural responses to output and factor price changes, and to other policy interventions, can be quite different from those suggested by conventional analysis. Hence, detailed empirical models of the agricultural household that recognize its dual role as a producer and consumer unit, in a theoretically consistent manner, become essential tools for policy analysis.

The methodological approach to analyzing farm household decision making that explicitly accounts for the inter-relationships between production, consumption and labour supply decisions is referred to as a farm (or agricultural) household model. The empirical application of this approach was pioneered in the late 1970's, building on the theoretical work on the "subjective equilibrium" of the family farm/household developed by Nakajima (1969) and Sen (1966), among others.[1] There is now a large literature on estimating farm household models for developing countries and on their empirical applications for a wide range of policy questions.[2]

A key feature of most empirical applications of farm household models is that the many kinds of human labour inputs observed in traditional agriculture are simply added together into an aggregate homogeneous

1

labour input category. The aggregate labour input is then assigned a "price" equivalent to the prevailing market wage rate on the local hired labour market. In some cases the relative productivity differences of labour categories (e.g. male and female, or adult and child) may be indirectly incorporated by creating an aggregate labour input with fixed conversion factors. For instance, when the labour input data is dis-aggregated by gender, a common technique is to create an aggregate labour input by specifying that one unit of female labour is equivalent to, say, 0.8 units of male labour. Such conversion factors can be *ad hoc* or can be based on the ratio of the observed market wage rates for male and female labour.

Another important dimension to dis-aggregating total farm labour input is the distinction between labour supplied by family members of the farm household and the labour input provided by labourers hired at a given wage rate. The distinction between *family* and *hired* labour in farm household models is relevant not just from an empirical perspective of capturing any productivity differences between these two types of labour inputs. This distinction is also important for methodological reasons because it determines the analytical structure of the farm household model, and the manner in which the household equilibrium can be determined. In the classification scheme adopted in Nakajima's classic treatise (1986) on farm household models, the ratio of family and hired labour use is one of the two key dimensions along which farm households can be classified.[3]

At one extreme are settings where no local hired labour market exists and all farm cultivation is done by family labour. In this situation, where the household is in a state of "autarchy" (i.e. neither buys nor sells any labour), the household labour supply and labour demand equilibrium is characterized by a "subjective equilibrium" (Nakajima, 1986). Such an equilibrium implicitly defines an unobserved "virtual" or "shadow" price of labour, which, if it existed in reality, would be the price (the wage rate) that would induce the household to equate the demand and supply for its own family labour, taking that implied market wage rate as exogenously given (Strauss, 1986: 77). The key analytical feature of an autarchic household is that its preferences over leisure and consumption of goods, together with the farm production function, jointly determine the optimal demand for labour in farm production. Optimal production and consumption plans are formulated simultaneously.

At the other extreme are farms that rely completely on non-family labour employed at a specified market wage rate. These are "farm-firms". Their decisions on labour allocation for farm cultivation (and other optimal input choices) are exactly analogous to the textbook version of the profit maximizing firm. The consumption choices of the farm-firm household can also be modelled as in the textbook version of a consumer household. It chooses an optimum consumption bundle, given its budget constraint (which in this case includes income from farm profit) and exogenously given market prices, including the market wage for hired labour as the appropriate price of leisure. The only special condition to take account of is the revealed preference for zero amounts of family labour supply. The non-labour participation of the household members indicates that the subjective valuation placed on family leisure must exceed the market wage rate at which family labour could be sold. Such non-participation kinks are readily handled in labour supply models (Pudney, 1989).

The vast majority of farms in both developing and developed countries fall into an intermediate category where both family and hired labour are used in farm cultivation. In addition, with active rural labour markets, farm households can choose to allocate family labour to own farm cultivation or to wage labour employment in the local off farm labour market.

The conventional approach to empirical estimation of a model for a farm household that uses both family and hired labour as inputs in farm production is to treat family and hired labour as homogeneous inputs that can be substituted for each other on a one for one basis. Total labour input on the farm is simply taken to be the sum of family and hired labour days. It is also widely assumed that the local wage rate at which labour can be hired is also the appropriate wage rate applicable to family labour. Such an imputation is often made even if family labour is fully devoted to own farm work and does not supply any labour on the hired labour market (as in Rosenzweig, 1980).

These twin assumptions about the equivalence of a unit of family and hired labour (homogeneity), and an inferred price for family labour set equal to the given market wage rate (exogeneity of price), have the effect of making family labour equivalent to a tradable input with a price that does not depend on the farm-family's own decisions. If all other farm inputs and outputs are also tradable commodities with exogenously given market prices that the farm household takes as given, then the analytical structure of the farm household model is considerably simplified. The production and consumption decisions need not be modelled *jointly* (which makes empirical estimation relatively difficult) but can be solved *recursively*, so that the production and consumption decisions can be determined independently.

In the recursive framework, the farm household utility maximization problem is solved in two stages. In the first stage, the household acts as a profit maximizing producer. It chooses optimal levels of farm inputs, including total labour (which is a homogeneous composite of family and hired labour), given the exogenous set of prices for all inputs and outputs that the household faces. The solution to this first step determines farm profits and the full income of the farm household. Given this income level, in the second stage the farm household acts as a pure consumer, choosing the optimal consumption bundle, including leisure, subject to the full income budget constraint that is conditioned on the optimal production choices. A simple interpretation of the recursive property of the farm household model is that the profit maximizing production input and output choices are independent of the household's utility function (Strauss, 1986).

An intuitive way to understand the recursive property is that it necessarily holds as long as there are markets for all commodities and inputs, and the household is a price taker in all these markets. Then the amount of, say, the food crop to be produced by the household can be determined independently of the preferred level of household food consumption. The household can always buy and sell any amount of food at the going market price. Similarly, household preferences about leisure relative to the other consumption goods, together with the market wage rate, determine labour supply independently of the labour requirement on the production side. Again, any difference between household labour supply and own farm labour demand can be costlessly equilibrated through hiring in outside labour or hiring out family labour at the going wage rate.

Although the treatment of family and hired labour as homogeneous inputs greatly facilitates empirical estimation of farm household models, there are important reasons for carefully assessing whether family and hired labour are indeed homogeneous inputs in a given setting.

The main reason why family and hired labour may differ in terms of efficiency units is due to the "principal-agent" type of relationship inherent in most hired labour contracts (Feder, 1985). Because of the difficulty inherent in monitoring the effort applied by hired labour with a limited stock of family labour available to the farm household, "shirking" on the effort applied by hired labour may be commonplace (Binswanger and Rosenzweig, 1986). On the other hand, family labour has an incentive to work intensively because on the owner operated family farm it is the residual claimant to output. Many of the institutional structures of traditional agriculture – for instance, multi-generation joint families, sharecropping contracts, piece-rate wages, labour exchanges – seem designed to get around the problem of monitoring the work effort of hired labour.[4] Hence there is implicit recognition that the efficiency differences between family and hired labour related to the effort applied in farm work can be large.

Apart from the implications for an analytical understanding and modelling of the labour demand and supply decisions of farm households, any observed efficiency difference or heterogeneity between family and hired labour can also have significant policy implications. If the efficiency differences between family and hired labour are related to work incentives, then aggregate outcomes – e.g. total production, total labour demand (labour absorption), and equilibrium rural wage rates – are sensitive to changes in the distribution of individual endowments of land and labour.

If, for instance, family labour is more productive because it applies more *effort* per unit time than hired labour, then a re-distributive land reform program which transferred land from big farms relying primarily on hired labour to small family-labour-operated farms would increase the average labour intensity of cultivation, and hence total absorption of labour in the agricultural sector. On the other hand, the *increase* in aggregate labour demand could be associated with a *reduction* in the demand and supply for hired labour such that the equilibrium market wage rate for hired labour could be reduced. This would adversely affect landless households that were not beneficiaries of the land transfers. The direction and exact magnitude of such general equilibrium effects depend on the precise values of the various elasticities of labour demand and supply for the different classes of farm households (Rosenzweig, 1978). Hence, from a policy perspective, it is useful to devise a suitable framework and methodology to estimate the parameters of a farm household model that allows for the heterogeneity between family and hired labour.

The empirical literature on testing whether family and hired labour are heterogeneous inputs, and, if so, what determines their relative efficiencies, is rather scant. Moreover, existing studies mainly look at the heterogeneity and efficiency issue solely from the production function perspective, without relating the heterogeneity to the utility maximizing labour/leisure trade off faced by the household in the presence of labour heterogeneity. Examples of the production function approach are Frisvold (1994), Deolalikar and Vijverberg (1987), and Bardhan (1973). These previous studies on the efficiency differences between family and hired labour do not specify an integrated farm household model structure where the labour supply implications of the observed heterogeneity have also been derived in a theoretically consistent manner and verified empirically. It is this shortcoming of the existing literature on labour heterogeneity that has motivated the approach taken in this book.

1.2 The Research Question

The specific research question addressed is whether, in the setting of Nepalese agriculture, family and hired labour are equivalent (or homogeneous) inputs in farm production. If not, what is the nature and extent of heterogeneity between family and hired labour? And how can one amend the conventional recursive farm household model structure to estimate a theoretically consistent labour supply equation for family labour, taking account of the specific form of the heterogeneity between family and hired labour detected empirically.

There are two distinct though related questions regarding family and hired labour being homogeneous inputs:

i.   Are family and hired labour perfect substitutes for each other in the production function? Is the elasticity of substitution between family and hired labour infinitely large? [5]
ii. If so, are the marginal products per unit of labour time of these two types of labour equal to each other, everything else held constant? [6]

If the answer to either question above is no, then family and hired labour become heterogeneous inputs. The total effective labour input on a farm utilizing both family and hired labour workdays will be something different than the simple sum of the labour days put in by family and hired workers. More importantly, with labour heterogeneity the wage rate for family labour that applies to the determination of the farm household's labour-leisure equilibrium will differ from the market wage rate paid for hired labour. With this shadow wage rate for family labour being unobservable, as well as possibly endogenous to the household's labour allocation decisions, the recursive or separable property of the farm household model is lost and estimation becomes cumbersome.

In the presence of labour heterogeneity, the analytical challenge is to seek ways in which separability could be restored in a theoretically consistent way by relating the shadow wage rate of family labour to the observed market wage rate for hired labour, taking into account the precise nature of the efficiency differences between family and hired labour. If the appropriate shadow wage rates for family labour can be recovered from the observed market wage rates and from the fixed parameters of the production function which describe the extent of the efficiency differences, then separability of the farm household model is restored. The labour supply component can be estimated with the shadow wage rate generated in this manner. Whether such an approach is feasible will depend on the type of the labour heterogeneity indicated in the data and on how the marginal products of hired and family labour behave.

An important limitation of the research reported in this book needs to be noted at the outset. It does not delve into the underlying causes or sources of the efficiency differences between family and hired labour implied by the production function estimation results presented in Chapter 6. This research is not intended to discriminate between alternative explanations for why such labour heterogeneity may arise – e.g. whether it is due to the principal-agent incentive problem for hired labour, or the farm-specific experience of elderly family members (Rosenzweig and Wolpin, 1986), or due to some undetected labour market imperfections. It would be worthwhile to pursue the underlying causes; but they are properly the subject matter for an altogether different book, and one that cannot be adequately supported by the type of survey data used in this study.

1.3 Estimation Methodology

The empirical estimation strategy adopted is based on a two step procedure suggested by H. Jacoby (1993) for estimating non-separable farm household models.[7] In the first step Jacoby estimates a farm production function with family labour as a distinct labour input; and in the second step a structural labour supply equation is estimated to represent the household's utility maximizing equilibrium over leisure and consumption of other goods. The labour supply equation is based on an unobservable shadow wage rate for family labour. Relying on the optimal first order conditions, Jacoby uses the estimates of the marginal product of family labour derived from the production function parameters to generate the unobservable shadow wage rates used as regressors in the second step.

Jacoby's prime interest is not to test for efficiency differences between family and hired labour. The same sequential approach, however, can be used to estimate a farm household model structure in which is embedded a test for the heterogeneity of family and hired labour. In the second step the labour supply functions are estimated in a theoretically consistent manner with the labour heterogeneity found in the production function estimation.

The heterogeneity of family and hired labour is tested through several alternative specifications of the production function that permit imperfect substitution between family and hired labour and differences in their marginal product. Rejecting the parametric restrictions that lead to a model with homogeneous labour is the statistical evidence for heterogeneity.

In the second step, some of the variables used in the labour supply regressions will be generated from parameters estimated in the production function.[8] The main variable of interest derived in this manner is the appropriate shadow wage rate that reflects the true opportunity cost of family labour at the equilibrium labour supply position. A key question that arises is how this shadow wage rate is related to the off farm market wage for family labour and the wage for hiring in labour on the farm.

The first order conditions for an optimal equilibrium in a farm household model with heterogeneous labour inputs can be used to relate the shadow wage rate to the observed market wage rates and the parameters that describe the extent of the labour heterogeneity. The labour supply equations can then be estimated with the appropriate shadow wage rates derived in this manner. The resulting labour supply regression parameters then describe the true labour/leisure choice of the farm household that is consistent with utility maximization in a setting where the specific type of labour heterogeneity modelled in the production function is observed.

1.4 The Setting

The data used in the empirical section of this book comes from a sample of farm households from the southern plain (or *tarai*) region of Nepal. The household level survey data used for the estimation are drawn as a sub-sample of a nationally representative Household Budget Survey conducted by Nepal Rastra Bank (the Nepalese central bank) in 1984/85. This survey had a farm management module in which data on output and input use was collected on a crop basis over a multi-round survey period. The data separately identifies inputs of family and hired labour, which are further broken down into male and female categories. This survey also included a module on family labour use by household members aged ten or above, with which labour supply estimation can be done at the individual level.

The estimation of the farm household model is based on a sample of approximately one thousand households drawn from five of the twenty districts of the southern plain *(tarai)* region of Nepal (see Figure 1.1). Data for a larger sample of households from the northern hill and mountain regions of Nepal are also available in this Nepal Rastra Bank survey. But these households have been excluded from this study because of the very limited use of hired labour in the northern hill and mountain regions of Nepal. Farm sizes in these northern regions are rather small and production is mainly subsistence oriented. There is only a limited form of economic differentiation among farm households in these northern regions.

The agrarian structure in the *tarai* region of Nepal closely follows the classical structure: most villages will have a few large landlords dependent primarily on hired labour, a middle group of owner-cultivators and a large group of landless households who supply the hired labour on the larger farms and may also typically rent in small plots of land. In the *tarai* region it becomes feasible to consistently distinguish between households that are net buyers and sellers of labour on the local village labour markets.

The rural labour markets in the *tarai* region are quite active. In peak periods there is also considerable in-migration of agricultural labourers from India. Rural households in the *tarai* region of Nepal are less likely to be constrained quantitatively on either the labour demand or labour supply side than farm households in the northern hill/mountain regions. Hence the modelling of efficiency differences between family and hired labour and the implied theoretically consistent labour supply behaviour is best addressed in the context of the *tarai* region only.[9]

The productivity of agricultural labour is a central theme in any policy discussion of rural development and poverty alleviation in Nepal. Agriculture still accounts for about 50% of GDP. About 90% of the Nepalese people live in rural areas and more than 80% of them depend on agriculture for their livelihood. Agricultural production technology is still very backward in Nepal. Human labour is the main farm input, typically accounting for more than 50% of the total cost of farm cultivation.[10]

The parametric estimation of a farm household model structure for Nepal with a specific focus on labour productivity is useful work in itself, apart from the methodological concerns of correctly modelling the farm household's production and consumption decisions under labour heterogeneity. There has been little quantitative work of this nature on Nepalese agriculture utilizing data from large sample surveys. Obtaining more precise estimates of these parameters – for instance, labour supply and factor demand elasticities – would be valuable inputs for the analyses of many agricultural policy issues in Nepal, including applications of general equilibrium simulation models.

1.5 Chapter Outline

Chapter 2 provides a brief discussion of the background issues related to the question of labour heterogeneity. It also reviews the literature on the methodology as well as the findings on the tests of the heterogeneity of family and hired labour reported in previous studies, and on related aspects

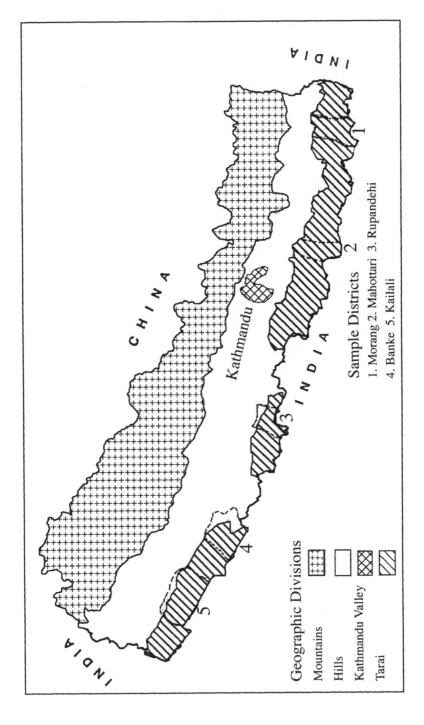

Figure 1.1   Geographic Divisions of Nepal and Sample Districts

of the larger literature on the labour supply behaviour of farm households that are relevant for this study.

Chapter 3 develops the analytical framework of a farm household model with heterogeneous labour inputs, and it also discusses briefly the implications of specific forms of heterogeneity on family labour supply. This chapter also shows how the correct specifications of the labour supply equations differ among households in the presence of labour heterogeneity.

Chapter 4 develops and discusses the specific econometric estimation methodology adopted in this book. The approach essentially follows the sequential two step procedure proposed by Jacoby (1993), with a specific adaptation that relates, in a theoretically consistent manner, the variables in the labour supply estimation to the type of labour heterogeneity detected in the production function estimation.

Chapter 5 briefly describes the Nepalese agriculture setting from which the survey data used in this book is drawn. It also provides an overview of the survey data itself and the derivations and definitions of variables used in the estimation chapters.

Chapter 6 contains the main estimation results on the tests for labour heterogeneity and alternative aggregator functions to create an effective labour composite, combining family and hired labour. It also presents the full estimates of the production function parameters and the elasticities of substitution and factor demands derived from these estimated parameters.

Chapter 7 presents the labour supply estimation results consistent with the nature of labour heterogeneity detected in Chapter 6. It also compares the model fit and parameter estimates of alternative specifications of the labour supply models with and without the heterogeneity adjustments.

Chapter 8 contains the summary and conclusions. It also briefly draws out some implications of the research and provides suggestions for further research work.

Notes

1   The initial theoretical reference to farm household models was a 1923 book by the Russian economist, A. V. Chayanov, of which an edited English version was published as Chayanov (1966). There were also several important non-English publications of Japanese economists – i.e. Nakajima (1949), Tanaka (1951). Early examples of the empirical estimation of farm household models are Lau, Lin and Yotopoulos (1978) for Taiwan, Kuroda and Yotopoulos (1978 and 1980) for Japan, Adulavidhaya, *et. al.* (1979) for Thailand and Barnum and Squire (1979) for Malaysia.

2   Two important collections of theoretical and applied work on farm household models can be found in the edited volumes Singh, Squire and Strauss (1986) and Caillavet, Guyomard and Lifran (1994).

3   The other dimension is the proportion of farm output consumed by the farm household. Various categories of farm households, with an underlying distinct analytical structure for the joint decisions on household production and consumption, result from their different exposure to the hired labour and farm output markets (Nakajima, 1986: Chapter 1).

4   In broader terms these arrangements in traditional agriculture can be seen as an institutional response to the high likelihood of incomplete or missing markets, particularly with respect to labour (de Janvry, Fafchamps and Sadoulet, 1991).

5   In a production function with more than two inputs there are many alternative measures of the degree of substitution between two specific inputs (Chambers, 1986: Chapter 1). A commonly used measure is the partial Allen Elasticity of Substitution (AES) which gives the effect on the quantity demanded of one factor due to a change in the price of another factor, holding output and other factors constant (Squires, 1994:186).   A positive value of the AES implies the two inputs are substitutes, or more formally, price or p-substitutes (Seidman, 1989). The higher the value of the AES the greater the degree of substitutability. For perfect substitutes, the AES is infinitely large. This implies a linear aggregate labour composite of the form Le = aF + bH, where Le is aggregate effective labour, F is family and H is hired labour, and a, b are constants.

6   Since marginal products of inputs vary with the levels of other inputs, the comparison of the marginal products of family and hired labour has to be made at some average level of all inputs in the sample of farm households. But if the production function is separable in the labour inputs, in which case the ratio of the marginal products of family and hired labour depend only on the levels of the labour inputs (Chambers, 1986: 43), the comparison of marginal products can be made in reference to the average level of the labour inputs only, ignoring all other inputs. See Chapter 3 for details.

7   Jacoby's approach is a general methodology for estimating a structural labour supply equation for workers who are self-employed. It is analogous to the treatment of labour supply in the presence of progressive income taxes that was pioneered by R. E. Hall (1973) and which involves "linearizing" an underlying non-linear budget constraint.

8   The sequential estimation of models that contain variables that are unobservable but can be estimated from an auxiliary statistical model is a standard procedure. Early examples from many areas of applied econometrics are cited in Murphy and Topel (1985). Pagan (1984) refers to these as models with "generated regressors". The sequential estimation strategy yields consistent estimates of second stage parameters under fairly general conditions, requiring only a simple adjustment of the standard errors for the second step equation (Pagan, 1986 and Murphy and Topel, 1985).

9   Comparative analyses of farm household behaviour and model parameters for the *tarai* and northern hill regions of Nepal would be of interest, but this has been left for subsequent research.

10 See HMG/N, Ministry of Agriculture, *Costs of Production for Major Crops in Nepal 1985/86* (Ministry of Agriculture, Kathmandu; 1987).

2 Background and Literature Review

This chapter has two main objectives. The first part contains a brief discussion of several themes in the development economics literature which draws attention to the question of heterogeneity between family and hired labour. The main themes explored are labour market dualism and the celebrated inverse relationship between farm size and land productivity. The second part reviews the literature that is directly related to the tests of labour heterogeneity and the implications of heterogeneity for the analytical structure and estimation of a farm household model. This literature review has three sub-sections: (a) a summary of the previous specification and results from production function based tests of heterogeneity between family and hired labour; (b) a brief review of the labour supply estimation research work that is relevant from the perspective of labour heterogeneity; and (c) a review of the Nepal-specific farm household model estimation and other relevant studies related to the empirical analysis of this study.

2.1 Background Issues

One of the key stylized facts about agricultural production in developing countries is that small farms are cultivated more intensively (i.e. with higher levels of variable inputs used per hectare) than bigger farms; and this often leads to an observed inverse relationship between farm size and productivity (output per hectare).[1] The relatively greater application of inputs on smaller farms is most pronounced for labour inputs. The labour input per hectare on small farms is consistently higher than on bigger farms over a large range of farm sizes; and this result holds whether or not average yields on small farms are higher (Berry and Cline, 1979).

Based on the above stylized facts of traditional agriculture, many authors went on to claim that a re-distributive land reform policy would be desirable not only from an equity point of view, but also from an efficiency consideration because breaking up hired-labour-based large farms into smaller family-labour-based farms would also increase agricultural output (Berry and Cline, 1979). Such a line of *ceteris paribus* reasoning, however, can be quite misleading. Everything else is unlikely to be the same after the land reforms. The early literature also did not address precisely what were the sources of the labour market dualism leading to higher intensity of cultivation on the smaller farms, and whether they would still be maintained after the land transfers.

Traditionally three general explanations have been offered for the observed inverse relationship, with regard to both labour input and yields:

i.   decreasing returns to scale;
ii. omitted variable bias due to unobserved higher quality of land on small farms; and
iii. factor market imperfections, especially in rural labour markets, that induce small farmers to apply too much family labour on their own farm cultivation, relative to the market wages they face.

Of these three explanations, the empirical evidence for decreasing returns to scale in traditional agriculture is quite weak (Bardhan, 1973; Moll, 1990). Also, it is not a plausible explanation in settings of extremely small average farm sizes. For instance, Benjamin (1995) documents a strong inverse relationship in Java within a sample of farms where the mean farm size is 0.71 hectares with a standard deviation of 0.02. It is unlikely that the technology available to farmers within such a narrow distribution of farm size would exhibit decreasing returns to scale.

More recent work on the inverse size and productivity relationship has focussed on land quality heterogeneity (Bhalla and Roy, 1988). This new evidence, especially from India, indicates that smaller farms tend to be more productive because of better irrigation facilities and other innate land quality differentials that in turn induce the higher labour input. In such a setting, breaking up the big farms into many smaller ones would not have any output increasing effect.

The evidence is mixed on whether land quality differences can account for most of the observed inverse relationship. Where the data allow for controlling the effects of possible correlation between farm size and unobserved land quality, the inverse relationship with respect to output sometimes disappears, as in Bhalla and Roy (1988) with Indian data, or still remains strong, as in Carter (1984), also with Indian data.

The more traditional explanation for the inverse relationship focussing on labour market imperfections was based on a hypothesized difference in the "mode of cultivation" between small and big farms, leading to a form of labour market dualism within the agricultural sector. The causes and consequences of labour market dualism in developing countries has been an important strand in the development economics literature for a long time. While inter-sectoral wage gaps (i.e. a gap between the rural agricultural labour wage rate and an urban sector labour wage rate in real terms) have been the main focus of this literature, following Harris and Todaro (1970), the presence of dualism within the rural/agricultural sector is also widely recognized (Mazumdar, 1975, 1989; Ahmed, 1981).

In a farm household model framework, the essence of labour market dualism lies in the simple presumption that the real cost of unpaid family labour – measured as the rate of utility substitution between leisure and consumption – is not equated to an observed market wage rate for hired labour (Sen, 1975). Consequently, the real cost of (identical) labour can differ among rural households even in a localized village labour market.

This line of explanation, while plausible in the institutional setting of traditional family based agriculture, is not wholly consistent with the empirical facts that small farms also rely to a considerable extent on hired labour; and that a large part of the hired labour on big farms is supplied by self-cultivating small farmers. In addition, there is growing empirical support for and general acceptance of relatively well functioning rural labour markets (Rosenzweig, 1988; Benjamin, 1992) which goes against the main grain of Sen's explanation. One should note that the existence of some form of labour market imperfections also implies that there are related imperfections in the land rental markets that prevent the wage gap being erased through changes in the scale of operation of the small and big farms (Binswanger and Rosenzweig, 1986).

An extreme form of the inverse relationship between farm size and productivity occurs when the output per hectare is strongly declining in the size of a plot within the same farm household. Udry, *et al.* (1995) estimate a farm production function using the very detailed plot level data collected by ICRISAT from Burkina Faso. With a panel data estimation framework they show that, controlling for household and annual fixed effects, output is strongly declining with the size of the plot within a farm. The usual factor market imperfection story cannot explain such a relationship since it occurs among plots cultivated within the same household, and not across households that may face different opportunity costs of labour. Among a variety of possible explanations, Udry, *et al.* note that their result may be due to the problem of monitoring the labour effort applied on the bigger plots within a household.

An alternative explanation for the lower labour input per unit of land on bigger farms, which does not rely on labour market failure, is to explicitly account for the differences in the efficiency of hired and family labour time due to difficulties in monitoring the effort applied by hired labour. Feder (1985) has formalized a model with supervisory costs of using hired labour (and with credit constraints) which gives rise to an inverse relationship between farm size and labour use and output, even with well functioning land and labour markets. Incentive related differences in the effort applied by farm labour in other forms of agricultural organizations, such as collective farms, has also been noted (Sicular, 1986).

However, there has been very limited work on empirical testing for heterogeneity between family and hired labour. Still less work has been done to show whether any observed heterogeneity can be related to the principal-agent type of incentive problem (Douglas, 1989), or to the related costs of monitoring the effort applied by hired labour (Binswanger and Rosenzweig, 1986). If labour heterogeneity is related to work incentives, with family labour being more productive, this would offer an alternative explanation for the inverse relationship between labour use and farm size without explicitly relying on factor market imperfections.

Even when there is clear evidence for labour heterogeneity, with family-labour-based small farms being more productive, the efficiency argument for land reform is not clearly established without knowing in detail the sources of the labour heterogeneity. It could be that family labour is more productive solely because the family members in households that own or operate some land are better fed than hired workers who usually come from landless households. There is some evidence that nutritional status does affect the effort expended in farm work (Deolalikar, 1988). If in a specific setting it turns out that nutritional deficiencies are the key source of any observed labour heterogeneity, then the incentive related analytical categories of family and hired labour are not relevant. The nutritional deficiencies of the landless workers could be remedied without turning them into owner-cultivators (although doing so through a re-distributive land reform program would also improve their nutritional status and hence productivity). The policy issue in such a situation then becomes one of choosing a first-best option. In general it is important to focus on the actual sources of the productivity advantages that small farms may enjoy and to clearly link policy options to the underlying causal factors.

2.2 Literature Review

This section provides a more detailed review of the specific methodology and results obtained in previous studies in relation to direct or indirect tests for heterogeneity between family and hired labour. A second sub-section reviews the labour supply estimation literature from the perspective of the implications of labour heterogeneity. A third part of this section reviews the Nepal-specific literature with respect to empirical estimation of farm household production and labour supply decisions.

2.2.1 Labour Heterogeneity

The main approach to testing for heterogeneity between family and hired labour has been through production function estimations that rely on alternative specifications of the labour input. Tests for labour heterogeneity in a production function framework ask two different but related questions. First, are family and hired labour perfect or imperfect substitutes in the farm production process as measured by the elasticity of substitution between them? Second, regardless of the degree of substitution between them, are the marginal effects on output from a unit increase in family labour the same as the effect from an unit increase in hired labour? The latter is a test for the equality of the marginal products of family and hired labour from a purely technological perspective.[2]

These tests can be done with family and hired labour specified as two distinct inputs in the production function (Jacoby, 1993 and Squires and Tabor, 1994). Or, the production function can have a nested structure in which, at an initial step, a composite labour services variable can be specified as a function of the observed levels of family and hired labour inputs. At a subsequent stage, the composite labour variable enters as a single input in the farm production function with other non-labour variables (Deolalikar and Vijverberg, 1983, 1987 and Frisvold, 1994).[3] The former is a more general specification, but it is not always feasible to implement, especially if there are many cases with zero-values for any of the labour categories in the sample data.

Bardhan (1973) is one of the earliest examples of a specific test for labour heterogeneity embedded in a production function estimated with farm level data (from the Indian Farm Management surveys). Bardhan's production function specification was Cobb-Douglas where the ratio of hired labour to total labour entered as a separate input in addition to total labour (measured as the sum of family and hired labour). The value and the sign of the coefficient on the labour ratio variable provide a test for labour heterogeneity. For most of his different samples Bardhan finds that the coefficient for the ratio variable is insignificant, implying homogeneous labour. In the few cases when the ratio variable is significant, its sign is positive, implying hired labour is more productive than family labour.

Deolalikar and Vijverberg (1983) have criticized Bardhan's test as being based on an unsatisfactory functional specification, and also for not distinguishing clearly between imperfect substitutability and differential marginal effects. In particular, a Cobb-Douglas specification with the ratio of hired labour to total labour entered as a separate variable implies that hired labour is an essential input. Output is zero if the hired labour ratio is zero; and this can possibly bias the coefficient on the ratio variable to be positive (Deolalikar and Vijverberg, 1983: 48).

Using aggregate district-level data for India, Deolalikar and Vijverberg (1983) used a more general functional specification to test for labour heterogeneity. They specify a nested production structure in which the overall production function is Cobb-Douglas with a composite labour services input. The composite labour input, in turn, is specified under various alternatives of a CES or Diewert's generalized linear production function. They find strong evidence for limited substitution between family and hired labour. Their estimates of the Allen partial elasticity of substitution (AES) between family and hired labour ranged from 0.6 to 2.4, under different specifications.[4] They also find substantial differences in the marginal product of family and hired labour, with family labour being more productive.

Using conventional *F* tests on the restrictions of the model parameters and the Davidson and MacKinnon (1981) *J* test for non-nested models, Deolalikar and Vijverberg conclude that their "best specification" of the composite labour input in the agricultural production function is a Cobb-Douglas relationship between family and hired labour. This specification actually transforms the entire production function also into a Cobb-Douglas relationship with family and hired labour as distinct inputs.[5] Such a preferred specification is due in part to the nature of the aggregate district level data in which all sample points have non-negative values of both the family and hired labour input. In their preferred specification the marginal product of family labour is 2.5 times the marginal product of hired labour at the mean of the data. While the finding of a higher productivity of family labour is plausible, such a large discrepancy in the relative marginal effect is implausible. If it is not due to the aggregate nature of the data, which the authors themselves cite as a possible explanation, it is a likely indication of the mis-specification of a Cobb-Douglas production function with family and hired labour as two distinct and independent inputs.[6]

Aware of the potential bias in the results due to the use of aggregate data, Deolalikar and Vijverberg (1987) made another effort to test for labour heterogeneity using household level data for a sample of Indian and Malaysian farms. The estimation methodology is similar, with a quadratic labour services function nested into a primary level Cobb-Douglas production function. Again, they find strong evidence for imperfect substitutability, but the result on the marginal effects are reversed, with an estimated higher productivity of hired labour in both the Indian and Malaysian samples. At the mean of the data, the estimated elasticity of substitution between family and hired labour is 0.68 in the Indian sample, and 1.16 in the Malaysian sample. In both cases the estimated values are not significantly different from unity, implying that a Cobb-Douglas specification for the composite labour input would be appropriate.[7]

The estimation results in Deolalikar and Vijverberg (1987) also show that the ratio of the marginal product of family to hired labour ranges from 0.32 in the Indian sample to 0.78 in the Malaysian sample (at the mean of the data). Even when family and hired labour are imperfect substitutes, it is difficult to explain why the marginal product of hired labour is three times the marginal product of family labour (as in the Indian sample). To reconcile this with profit maximizing behaviour the relevant wage rates for these two inputs should vary by a factor of three, which is a very implausible scenario.

Squires and Tabor (1994) use a translog production function specification to test for the degree of substitution between family and hired labour in a large, regionally stratified sample of Indonesian farms. Their estimates of the Hicksian elasticities of substitution[8] between family and hired labour fluctuate in sign, indicating a substitute relationship in some regions and crops, and a complementary relationship in others. Their overall results show a limited extent of substitution between family and hired labour, in particular when measured by the direct elasticity of substitution between them. These values are usually much smaller than one (the implied value from a Cobb-Douglas specification).

Frisvold (1994) approaches the question of labour heterogeneity from the perspective of supervision costs imposed by hired labour. Using a particular interpretation of supervision costs by assuming that monitoring of hired labour is more effective when family members work along side with hired labour, Frisvold measures the effect of employer supervision on the productivity of hired labour. If family and hired labour were perfect substitutes in production with no "shirking" behaviour by hired labour, the supervision effect should be insignificant. If supervision effects were present then family and hired labour would be imperfect substitutes with a certain degree of complementarity. Increased allocation of family labour, working along side with hired labour and providing supervision, would increase the marginal productivity of hired labour. Using the very detailed plot level production data collected by ICRISAT for their village level studies from southern India, Frisvold reports three major findings:

i.   family and hired labour are imperfect substitutes, and the elasticity of substitution decreases as the ratio of hired to family labour increases;
ii.   the productivity of hired labour increases with the supervision intensity of family labour, but at a decreasing rate; and
iii. the labour effort expended per unit of time of family labour is higher than that of hired labour, but this differential diminishes with increases in supervision intensity. At high levels of supervision, the productivity of hired labour approaches that of family labour.

Because of the very detailed ICRISAT data set used in Frisvold's study, these results are quite significant.

The question of labour heterogeneity has also been addressed in the context of other labour categories, particularly between the labour input of male and female workers.

Laufer (1985), using the same ICRISAT data for India, has estimated the elasticity of substitution between male and female labour based on a generalized quadratic production function with male and female labour as distinct inputs. She finds evidence for imperfect substitution, but her results, based on separate estimations for different crops, give widely divergent values for the AES. Male and female labour are estimated to be complements (negative AES) in rice and sorghum production but substitutes (positive AES) in legumes. Laufer explains the wide divergence of results by noting that the notional categories of male and female labour may actually be capturing the functional nature of the different tasks performed by male and female workers, given the strict division of labour by gender observed in her sample.

Another important result presented by Laufer is the comparison of the estimated marginal product of male and female labour. The marginal product of female labour is always less than that of male labour. Their ratio ranges from 0.49 to 0.77, and this pattern is consistent with the observed range of the female to male worker's wage ratio in the sample villages.

Udry (1996), using detailed plot level ICRISAT data for Burkina Faso, also finds substantial gender differences of a more general nature. Using a nested CES production function structure with a multi-factor CES labour nest that specifies four labour categories (adult family male, adult family female, family child and non-household labour), Udry's estimate of the elasticity of substitution for each pair of labour inputs is 2.3.[9] Udry also presents dramatic evidence that in a given household, where some plots are under the control of the husband and other plots under the control of his wives, the productivity on plots controlled by women is significantly lower than on plots controlled by the husband, even for the same crop. This result is not due to differences in technology or lower efficiency of female labour (which is concentrated on the plots controlled by women), but to an overall lower labour intensity of cultivation of plots controlled by women. Using the same data set, Udry, *et al.* (1995) actually find that the marginal product of female labour is higher than male labour over all plots.[10]

Empirical tests for labour heterogeneity have not always rejected the traditional assumption of homogeneity between family and hired labour. Benjamin (1992), using detailed SUSENAS data for Indonesia and only focussing on rice growing farms in Java, tests for a linear specification of an effective labour composite of the form $Le = L^F + \theta L^H$.[11] He does not reject the null hypothesis of $\theta = 1$, and hence does not reject the joint null hypothesis of perfect substitution and equal efficiency between family and hired labour. His tests also support a linear procedure for aggregating male and female labour, subject to a relative wage adjustment factor.

Elizabeth Field (1988) provides an interesting test for the heterogeneity between free and slave labour in the cotton plantation economy of the ante-bellum southern United States. Using a translog production function with free and slave labour as distinct inputs, she finds that the Hicksian elasticity of complementarity (HEC) between free and slave labour was positive on both small and large plantations. These two labour inputs were *q*-complements, in that a higher application of one increased the marginal product of the other.[12] She also tests for the separability of these two labour inputs from the other inputs in the production function and finds evidence for weak separability. Thus free and slave labour can be combined into some aggregate index of labour input; but a simple linear combination, implying that they were perfect substitutes, is not supported by the data.

Pitt and Rosenzweig (1986) provide a novel indirect test of labour heterogeneity and the underlying assumption of the recursiveness of the production and consumption decisions of farm households. Using household level data from Indonesia that identified the incidence and severity of ill health suffered by household members, they test whether the reported bouts of illness significantly affect the household's labour supply as well as farm profits. They find that while illness significantly reduced the labour supply, farm profits are unaffected. This is an indirect test for labour homogeneity: market substitutes apparently can be found for the significant illness-induced reduction in the farm family's own labour supply on farm cultivation. They interpret such a result as proof that family and hired labour are homogeneous inputs, and also as evidence for the recursive structure of the production and consumption/labour supply components of the farm household model.

2.2.2 Labour Supply

The labour supply decisions of farm households have not yet been as extensively analyzed as labour supply functions of workers in developed countries; and reliable estimates of the various elasticities of labour supply are few. But the available evidence (Singh, Squire and Strauss, 1986c1; Bardhan, 1979; Rosenzweig, 1980) show that labour supply responses of farm households are not uniform across different asset class and farm size variables. The effects of changes in the wage rate and asset income on the labour supply of individuals varies particularly between landless/small farm cultivators (who work mainly on the off farm labour market) and larger farm cultivators (who may allocate their total labour supply both to own farm cultivation and to off farm wage employment).

One of the main methodological issues in modelling the labour supply behaviour of farm households that devote some family labour to own farm cultivation is the identification of the wage rate to represent the true opportunity cost of family labour at the equilibrium position. There are three general approaches taken on this issue.

The first and most common approach relates the opportunity cost of family labour in all households to the observed off farm market wage rate, irrespective of whether a particular individual actually works off farm or not. This assumption is consistent with a recursive structure of a farm household model with homogeneous labour, where the labour supply decision can be modelled separately from the production decisions. This is the approach adopted in Bardhan (1979) and Rosenzweig (1980) which were some of the earliest examples of labour supply estimation for developing countries using large data sets. Assigning a common observed market wage rate as the opportunity cost of family labour is not always valid even under the assumption of homogeneous labour. (See the subsequent discussion in Chapter 3). This assumption is even more suspect if family and hired labour are allowed to be heterogeneous inputs.

A second approach to labour supply estimation for farm households in a developing country setting has been implemented by Jacoby (1993) and Skoufias (1994), where the opportunity cost for family labour is equated to an estimated value of the marginal product of labour in own farm cultivation. This is an adaptation of the general methodology for estimating a structural labour supply equation for self-employed workers when applied to a farm setting. It is analogous to the treatment of labour supply in the presence of progressive income taxes which relies on "linearizing" the budget constraint, as suggested by Hall (1973). In this method the estimation is done in a two step procedure. In the first step, a farm level production function is estimated, conditioning on the optimal levels of family labour supply that is observed. From the estimated parameters of the production function the marginal product of family labour can be derived for each sample household. These estimated marginal products are the appropriate shadow wage for valuing family labour and are also used to derive the shadow profit of the farm household from land ownership, which is included as non-labour income in the labour supply equation. In the second step, the labour-leisure equilibrium choice can be modelled as if the farm household was a pure consumer household which faced a parametrically given exogenous wage and non-labour income.

The third approach has been to jointly estimate the production and labour supply components of the farm household model in a non-recursive structure, allowing for the shadow wage rate for family labour to be endogenously determined. This approach, however, is analytically cumbersome and requires complex estimation methods. Examples are Lopez (1984) and Newman and Gertler (1994).

The data set used in Bardhan (1979) is a large sample of nearly 5,000 rural households in West Bengal from the Indian National Sample Survey. For the sample of male agricultural workers the wage coefficient in the labour supply equation is significantly positive. With the same model specification, Bardhan finds an insignificant wage response for male labour supply in the sample of own farm cultivators. The wage elasticity for male agricultural workers was estimated to be between 0.2 to 0.3. The wage response was particularly insensitive for female workers in both types of households, with some indication of a backward bending labour supply curve for female labour supply. Other variables such as caste and marital status appear to be more important determinants of female labour supply.

Bardhan (1984) repeats the labour supply specification of his 1979 paper with another sample of households in West Bengal from a different round of the National Sample Survey. In this sample the wage effect is negative for male labour supply, which is opposite of the result obtained in the earlier paper. The wage effect is positive but insignificant for female labour supply. In both studies Bardhan develops a variety of wage measures based on the reported market wage to accurately reflect the opportunity cost of family labour applied to own farm cultivation or the off farm market, but the wage sensitivity is quite low in most specifications.

Rosenzweig (1980), also using data from India and a labour supply equation specified with both male and female village level wage rates, finds that the supply of male labour is strongly backward bending for both landed and landless households. The labour supply curve for female workers is positively sloped with respect to the own wage, and the cross wage effect is negative and significant.

Jacoby (1993) applies the sequential estimation method to peasant farms in the highland region of Peru where family based subsistence farming is concentrated. In the production function estimation Jacoby specifies three labour inputs – family male, family female and total hired. Jacoby tests for the separability of male and female family labour from the other inputs and finds the underlying production technology exhibits non-separability. Hence a translog function, with different interaction terms for the three distinct labour inputs, is specified and from which the marginal products of male and female family labour are estimated. This specification, however, results in negative values of the estimated marginal product of family labour, particularly for female labour, in nearly 20% of the sample. Jacoby is forced to drop these observations from the labour supply estimation step. For the subset of the sample with positive marginal products, the mean value of the marginal product of female labour is about 60% less than that of male family labour.

The labour supply estimation results in Jacoby (1993), using an instrumented version of the computed marginal products as the shadow wage rate for the equilibrium allocation of family labour, are reasonable. Uncompensated own wage elasticities are positive for both male and female workers and the income elasticities are negative. This leads to significantly positive compensated own wage elasticities, which is consistent with utility maximization. The cross wage effects in both the male and female labour supply equations are positive (implying male and female leisure are complements in the household utility function); but these cross wage effects are small relative to own wage effects. Male own wage elasticities are higher than for female labour supply, which is counter to the usual result of higher own wage responsiveness of female labour supply, at least in developed country settings (Killingsworth and Heckman, 1986). Jacoby indicates this result is mainly due to the definition of labour supply he has used, which includes domestic household work. Such a definition of female labour supply is likely to have a lower wage elasticity.

Skoufias (1994) has applied the same methodology used by Jacoby to the ICRISAT data for India, in a panel data framework with fixed effects. He uses a Cobb-Douglas production function with four different labour inputs, distinguished by gender and family and hired source. While the fit of the estimated production function is very reasonable (the R-square is 0.92 in the fixed effects model) the computed values of the marginal products of labour at the mean of the data provide very divergent results for the productivity of the four categories of labour. For instance, the marginal product of female family labour is estimated to be 4.7 times the marginal product of hired female labour. Similarly, the marginal product of female family labour is more than double the marginal product of male family labour.[13] There are no clear reasons to expect such large divergence in the marginal products by gender at the production optimum point. This result is likely to be a problem of the Cobb-Douglas specification with the four labour variables as distinct inputs.[14]

The labour supply results in Skoufias (1994) are also not fully satisfactory, especially for female family workers. Using the instrumented values of the estimated marginal products of family labour as the shadow wage rates, the uncompensated own wage effect is positive in the male labour supply equation. But the cross wage effect (with respect to the (female wage) is negative, indicating male and female leisure are gross substitutes in this data set. The income effect is significantly negative for male labour supply, which is consistent with leisure being a normal good. Skoufias however obtains unusual results for the female labour supply equation – the gross uncompensated effect is negative, suggesting a backward bending labour supply curve. He offers the same explanation as Jacoby (1993) – that the labour supply variable includes all work activities, including housework. But alternative estimates for labour supply excluding housework are not provided.

A backward bending labour supply relation can, of course, be a valid result for the gross uncompensated own wage effect. But in Skoufias's results, because of the insignificant value of the negative income effect for female labour supply, even the compensated own wage effect is negative. This is inconsistent with utility maximization. These anomalous results are likely to have been caused by the widely divergent estimates of the marginal products of labour from his Cobb-Douglas production function specification with four distinct labour inputs.

Lambert and Magnac (1994) carry out only the first step estimation of the production function for a farm household model with distinct labour inputs; but they provide direct comparisons of the estimated marginal products and the market wage rates. Their results also indicate the problems inherent in deriving the marginal product of labour with many categories of labour defined in the production function. Using data for the Ivory Coast, they estimate a generalized linear production function and compare the estimates of the shadow price (marginal product) of labour with the market wage rates. Their production function specifies three different labour inputs: hired labour, family male and family female. Their estimates reveal large differences between the marginal product of these three labour inputs, which, while indicative of labour heterogeneity, are puzzling. For instance, in one set of estimates using instrumental variables, the marginal products of labour per day are estimated to be 157.2, 0.62 and 3.14 for hired, male family and female family labour, respectively.[15] These large differences remain unexplained and are even more puzzling given that the average market wage rate reported in their data is 5.88. Again, this is likely to be an indication of mis-specification of the production function with these three different labour input categories.

Given that Jacoby, Skoufias, and Lambert and Magnac all find a very weak correlation between the estimated marginal products and observed market wage rates, Strauss and Duncan (1995) question the validity of using the estimated marginal products as proxies for the shadow wage rate, particularly if the estimation is based on survey data that does not have a detailed farm management component.

Newman and Gertler (1994) develop and apply to Peruvian data the more general joint methodology for estimating the labour supply behaviour of self-employed family members in a non-recursive framework. They specify three different equations: a direct specification of the marginal rate of substitution (MRS) between leisure and consumption to reflect household preferences; a market wage function for individuals in each of several age-sex categories; and a marginal return to farm work function for individual by each age-sex categories.[16] This last equation is specified in terms of prices and quasi-fixed inputs and human capital related variables. It can be derived as the derivative of an implicit farm profit function with respect to farm labour for an individual in a specific age-sex group, conditional on the total application of family labour on the farm. This procedure allows them to model the heterogeneity in family labour across specific age-sex groupings both in the production and consumption components of the model.

On the consumption side, the MRS between leisure and consumption is, of course, not directly observable. But certain imputations can be made based on the first order conditions that equate the MRS to the marginal returns to labour. For individuals who report working in the off farm labour market, the wage they receive will be equated to the MRS. For individuals who work only on the family farm, the shadow wage, which is equated to the MRS, can be computed from the marginal returns function. Since there is a dependence in both the underlying profit and utility functions of one category of family labour on other categories, and since corner solutions for non-labour participation have to be explicitly modelled, this method of estimating the farm household model involves a great deal of econometric complexity, especially as household size and labour categories are increased. No other application of this approach appears to have been published to date, testifying to the great difficulty in implementing this methodology.

2.2.3 Nepal-Specific Literature

The production component of farm household activities in Nepal have been studied in a growing number of empirical studies, based on production and profit function estimation strategies. But the specific question of heterogeneity between family and hired labour has not been directly addressed before. The labour supply and consumer demand system estimation for Nepalese rural households have been relatively neglected. When it comes to an integrated farm household model estimation strategy, this author is aware of only two other studies, Acharya (1987) and Abdulai and Regmi (2000), which have used a theoretically consistent framework.

Agricultural production function based studies at the household level in Nepal reveal a great deal of divergence in the regional pattern of production and input use both between the northern hill and southern lowland (*tarai*) region, as well as among the five development administrative regions which are defined on a east to west basis (NRB/ADB, 1994). This variation indicates that in any large sample, with nationally representative data, regional fixed effects on productivity and labour use patterns are going to be important factors to consider.

Because of a higher man-land ratio, the northern hill area agriculture is labour intensive, and several studies have estimated rather low marginal productivity of labour in hill agriculture – i.e. Belbase, *et. al.* (1985). Similar direct estimates of the marginal product of labour for the *tarai* region agriculture are not available. Belbase and Grabowski (1985) and Grabowski and Belbase (1986) show divergences in technical efficiency of farms varying according to farm size; but they do not discuss the underlying causes.[17] The issue of the work intensity of hired labour and the more general question of heterogeneity between different labour categories in Nepalese agriculture have not been adequately addressed.

Mudhbary (1988) estimates an AIDS consumer demand system focussed only on the demand for various food categories. Labour supply/leisure demand is not considered. Several other studies look at labour supply in the limited sense of off farm employment, ignoring the labour supplied on own farm cultivation. An example is Rauniyar (1986) who provides labour supply estimation results only for the off farm workdays from a small household sample drawn from two districts in Nepal. He reports significant positive own wage coefficients for the labour supply behaviour of both men and women. As expected the incidence of off farm labour supply is negatively related to farm size.

The farm production component in Acharya (1987) is modelled with a quadratic profit function from which the first order conditions for the optimal factor demands provide the estimating equations. Acharya treats family labour as a quasi-fixed input and estimates only the demand for hired labour conditional on the available stock of family workers. She finds that the demand for hired labour is negatively correlated to the own wage for both male and female labour. The cross wage elasticities suggest male and female labour inputs are complements – an increase in the male wage rate also reduces the demand for female hired labour. She attributes this result to the strict gender based division of labour so that an increase in any wage reduces the demand for all types of labour.

On the labour supply estimation component, Acharya estimates labour force participation and days of work equations separately for male and female household members. The dependant variable in the labour supply equation is not a scale variable that reflects the actual amount of time spent on specific work activities. The dependant variable was assembled from time allocation data that recorded only the frequency with which an individual engaged in specific activities within the observation period. The actual duration of each activity was not recorded. Even with such a limited proxy variable for labour supply, Acharya reports very reasonable results. Both the labour market participation decision and the frequency of work variables have positive wage effects and negative income effects.

Apart from wage and non-labour income, Acharya's results indicate that ethnic and caste grouping and demographic characteristics, such as family size and the number of adult workers per household also affect individual level labour supply decisions.

Abdulai and Regmi (2000) have applied the sequential strategy of Jacoby (1993) to estimate family labour supply behavior in a sample of 250 farm households in Nepal. They first estimate a Cobb-Douglas production function with family and hired labour as distinct inputs, and use the computed marginal products of family labour (in an instrumented form) as shadow wages for estimating the labour supply equations. Their labour supply results show that for both male and female workers the own wage effect is significantly positive and income effects significantly negative.

Abdulai and Regmi's labour supply estimation results appear reasonable but their robustness must be questioned because, by not directly addressing the nature and extent of labor heterogeneity between family and hired labour in their sample of Nepalese farms, the shadow wages they derive from their estimates of the marginal product of family labour are extremely low compared to the marginal product of hired labour as well as the observed market wage rate.[18] This is likely to indicate mis-specification of the production technology with family and hired labour as distinct inputs in the production function. Their results reinforce Strauss and Duncan's (1995) misgivings about the usefulness of Jacoby's shadow wage methodology which tends to produce unreasonable estimates (including negative values) of the marginal product of family labour at the household level. Such errors seem magnified when family labour is treated as a fully distinct input in the presence of hired labour use, as in Abdulai and Regmi.

Cooke (1998) estimates the demand for labour in the hill region agriculture of Nepal, using the observed market wage rate to value family labour. The main focus of Cooke's analysis is to determine how labour allocation on the family farm is affected by the scarcity of environmental goods – such as firewood, fodder and drinking water – that rural households have to collect with family labour, especially of women. She estimates a labour demand system that includes as regressors the wage rates for male and female labour, as well as the shadow price of the environmental goods, which are the wage-based valuation of the time taken to collect them. She assumes that male and female labour inputs are imperfect substitutes in production, but within each gender family and hired labour are perfect substitutes. With this specification she finds that the demand for labour is either insensitive to the human labour wage or increasing with the wage.[19] The latter result is theoretically inconsistent.

Hence, the market wage may not accurately reflect the opportunity cost of family labour on the demand side as well.

The studies noted above provide a scattering of empirical work on rural households in Nepal. A complete farm household estimation strategy that also considers the question of labour heterogeneity between family and hired labour has not been carried out to date.

2.3 Summary

While there is a large and growing literature on the estimation of farm household models for developing country settings, the specific issue of labour heterogeneity between family an hired labour has not been addressed in an integrated farm household model framework. There are a few studies that have looked at labour heterogeneity purely from a production function perspective. Production function estimates reveal a low degree of substitution between family and hired labour as well as different marginal effects on output; but the results are mixed and affected by the tendency to use limited functional form specifications. The specific sources of heterogeneity are not clearly specified.

Attempts to estimate a non-recursive model with many different types of labour inputs lead to puzzling results with very large discrepancies in the marginal product of the different labour categories. Specifications based on an aggregate labour composite, allowing for heterogeneity between the different components, are likely to do better.

On the labour supply component, the approach most commonly used in the past (based on assigning the market wage rates to all farm households) produces mixed results. While more elaborate approaches to modelling the self-employment of family labour in peasant farming have been developed, these have not directly related the shadow wage rate applicable for family labour to the nature of the heterogeneity between family and hired labour. Nor do they take account of the fact that shadow wage rates should vary according to the labour hiring status of the farm household.

Alternative strategies of estimating labour heterogeneity and labour supply in a consistent manner have not been satisfactorily addressed in previous studies in general, and much less so in the context of Nepalese agriculture. In the case of Nepal, even ignoring the labour heterogeneity issue, there has been a very limited focus on the empirical modelling of the decision-making framework of rural farm households.

Notes

1   There is a large literature on this topic and it is not feasible to review it adequately in this chapter. The older literature is surveyed in Sen (1975) and Ghose (1979) mainly from an India-specific focus, and also in Berry and Cline (1979) with a Latin American focus. The size and productivity relationship has been clearly established in all other major South Asian countries: Bangladesh (Abedin and Bose, 1988), Nepal (Grabowski and Belbase, 1990), Pakistan (Mahmood and Haque, 1981). The inverse relationship is also robust to newer approaches to estimating farm production functions, including the random coefficient method (Hoque, 1988). For a dissenting view on the genuine nature of the inverse relationship see Rudra (1992).

2   This is a measure of the *technical efficiency* of hired and family labour. If there is a difference in the wage rates to be paid to family and hired labour, then at the optimal labour allocation, the ratio of the marginal product of family and hired labour would, of course, reflect the difference in their relative wage rates, if the farm household also achieved *allocative efficiency*.

3   Testing for labour heterogeneity with a nested production function structure is valid only when the main production function is separable in the labour inputs. Otherwise a consistent labour aggregate is not defined (Chambers, 1986; Berndt and Christensen, 1974). Empirical applications of the nested production function structure have not always been based on the required prior test for labour input separability.

4   These are the estimates of AES between family and hired labour in the function for the nested labour input. Such an AES is distinct from the AES between family and hired labour in the overall production function (Deolalikar and Vijverberg, 1983).

5   A Cobb-Douglas labour service "production function" nested within an overall Cobb-Douglas production function specified in terms of a composite labour variable is equivalent to a single equation Cobb-Douglas specification with family and hired labour as distinct inputs, without any additional parameter restrictions.

6   The data used by Deolalikar and Vijverberg (1983) for the labour input are not the actual hours of work but the average number of persons per farm holding in the family labour and hired labour category in a district. If the average hours actually worked for each labour category are different then the estimates of the marginal products of labour per hour will differ. If the number of hours worked on average by a family worker is higher than the average number of hours of a hired labourer, as is to be expected in poor family farms, then the ratio of the actual marginal product per hour of family and hired labour could be substantially less than the 2.5 value implied in their estimates.

7   Deolalikar and Vijverberg (1987) do not actually estimate a Cobb-Douglas labour nest because of the problem of zero values for hired labour in a large portion of the Indian sample. Ignoring this problem, the result of a unitary elasticity of substitution means the generalized quadratic labour services equation could be replaced with the Cobb-Douglas functional form.

8   The Hicksian elasticity of substitution and related concepts of input substitutability in a multi-factor setting are defined and discussed in Section 6.3.4 of Chapter 6.

9   Udry (1996: 1038). In a multi-factor CES production function the Allen partial elasticity of substitution between each pair of inputs must be the same (Uzawa, 1962).

10 Udry, *et al.* (1995) acknowledge the higher marginal product of female labour is due partly to crop-composition effects since women tend to specialize in high value vegetable crops grown on their plot with predominantly female labour.

11   Le is effective labour, L^F is family labour and L^H is hired labour, with θ representing a constant difference in their productivity in effective units.

12   The definition and interpretation of the Hicksian elasticity of complementarity is discussed more fully in Chapter 6 of this book, together with the need to mind the p's and q's when discussing substitutes and complements, as reminded by Seidman (1989).

13   Skoufias (1994) does not directly report the estimated marginal products of the four labour categories. They can be computed from the estimated parameter values for the production function and the summary of the data reported in his Table 1. The marginal products (per hour), at the mean of the data, are Rs. 0.87 for family male labour, Rs. 2.19 for family female, Rs. 0.27 for hired male and Rs. 0.47 for hired female categories.

14   As reported above in Section 2.2.1 of the literature review section on labour heterogeneity, Laufer (1985), using the same ICRISAT data set as Skoufias, reported a lower marginal product of female labour for all her crop-specific equations for the generalized quadratic production function. She confirms that her results are consistent with the observed lower market wages rate for female labour.

15   The units for the marginal products are not clearly reported since the output variable is an index. Since the authors directly test for the equality of marginal products and market wages, presumably the former are also in hundreds of the Central African Franc (CFA) in which the market wage rates are reported (Lambert and Magnac, 1994: 23).

16   Newman and Gertler's estimation procedure could actually be done with a separate marginal return to labour and MRS function for each individual, conditional on the labour supply behaviour of other household members. But this will only increase the already inordinate amount of time taken to iteratively solve and estimate their model. They report that a single run of their model took all night on a 486 personal computer.

17   Moll (1990) has raised a strong objection to the findings of Grabowski and Belbase (1986), claiming their results of lower technical efficiency on large farms is due to the curve fitting nature of their estimated production function, and not due to a test of the hypothesis of lower efficiency on larger farms. See Grabowski (1990) for a response.

18   Abdulai and Regmi (2000) report the mean of the estimated marginal product of family labour (averaged over all sample households) as Rs. 0.47 and Rs. 0.33 for male and female workers, respectively (Table 2, IV estimates); while the mean wage rate is reported as Rs. 35 for male labour and Rs. 30 for female labour (Table 1). This large discrepancy of average marginal products exceeding reported wage rates by factors of 75 and 90, respectively, is unreasonable, even allowing for some error in the units of reporting of these numbers. Similarly, when evaluated at the mean of the sample data, as given in Table 1, the estimated ratio of the marginal product of hired labour to that of family labour is 12.3 for male workers, and 10.6 for female workers. These again seem to be unreasonably high (and in this case are independent of any errors in the units of measurement).

19   In Cooke's estimation results the coefficient on the human wage rate variable will by definition be affected by the presence of the shadow price variables for the environmental goods. These shadow prices are just the valuation of the time taken to collect these goods, using the market wage rate to value one unit of time. So there is likely to be a strong correlation between these price variables. A separate labour demand equation without the other shadow wage variables was not reported. With respect to her main hypothesis, Cooke finds that labour input in farm cultivation is unaffected by the time taken to collect environment goods as reflected by the insignificant coefficients on the shadow prices for environmental goods in the labour demand equation.

3 A Farm Household Model with Heterogeneous Labour Inputs

3.1 Introduction

This chapter presents the analytical structure of a farm household model that allows for family and hired labour to be heterogeneous inputs in production. The model structure permits different types of heterogeneity, and homogeneity of the labour inputs occurs as a special case of the general model. The focus of this chapter is not on the specific sources of the heterogeneity, but the manner in which it affects the analytical structure of the basic farm household model. A key objective is to show under what conditions the recursive property of farm household models is maintained even with heterogeneous labour inputs; and to derive the implications of this structure for estimating the labour supply component of the model.

Section 3.2 presents a model where family and hired labour are treated as completely separate inputs, which is the most general form of labour heterogeneity. The next section discusses the several difficulties in specifying and in estimating a farm household model of this general form. Section 3.4 presents an alternative model with a nested production structure where, in an initial stage, family and hired labour are combined to create an aggregate or composite labour unit which then enters into the primary level farm production function. Section 3.5 derives the labour supply implications of the nested production structure with heterogeneous labour. One important advantage of a farm household model structure with a nested aggregate labour input is that the shadow wage rate for family labour can be related to the observed market wage rate for hired labour, when both family and hired labour are simultaneously used on the family farm. Such a relationship helps in identifying the correct effective wage rate to be used in estimating the labour supply functions for family labour even when it is applied solely to the family farm.

The structure of the farm household model presented in this chapter focuses on the treatment of family and hired labour as heterogeneous inputs. Many other important and analytically interesting issues that can be incorporated into farm household models (i.e. marketable surplus of farm output, production of non-agricultural goods) are not addressed in the models presented in this chapter. Our approach also ignores the Beckerian new household economics tradition of "home production" of consumption goods based on inputs of purchased goods and family labour time (Becker, 1965).[1] The models presented also adopt the convention of treating a multi-person farm household as a single decision making unit that maximizes household level utility, ignoring decision making based on individual utility maximization and bargaining among the household members.[2]

3.2 Analytical Structure of a Model with Heterogeneous Labour

Let U represent an aggregate household utility function defined over a composite consumption good C and leisure $\mathcal{L}$. Let T be the total labour time endowment of the household. T has three components: own farm labour use (F), market wage employment (M) and leisure ($\mathcal{L}$). The inputs in the farm production process are land area cultivated (A), family labour (F) and hired labour (H); and the two types of labour are initially treated as distinct inputs. For simplicity assume there is a single (or composite) farm output. Other variable inputs in the production process are ignored because they have limited bearing on the treatment of the heterogeneity between family and hired labour inputs.[3] The farm household also has non-labour endowment income E.

Let p represent the price of farm output and p_c (normalized to 1) the price of the composite consumption good. Let w^h be the wage rate for hiring in farm workers, and w the net wage rate that family labour receives when supplied in the hired labour market. (It is common to find $w^h > w$ because there likely will be some costs associated with working away from one's home; but no restrictions on these wage rates are imposed *a priori*). For simplicity assume that the net wage rate received for all off farm work is uniform – i.e. w applies to all alternative sources of employment for family labour outside of its own farm cultivation, whether it be agricultural work on big farms or non-agricultural work within or outside their village. It is assumed that labour markets clear and households do not face binding quantity constraints on either their labour demand or labour supply. The normal seasonal fluctuation in agricultural wages is also ignored; but the analytical framework of the model with a uniform wage will be valid in each of the specific periods during which the seasonal wage is fixed.

The household's utility maximization problem can then be set up as:

(3.1)   max $U(C, \mathcal{L})$

subject to

(3.2) $C = pQ + wM - w^hH + E$
(3.3) $Q = f(F, H, A)$
(3.4) $T = \mathcal{L} + F + M$
(3.5) $F \geq 0;\ M \geq 0;\ H \geq 0.$

The constraints 3.2 to 3.4 deal, respectively, with the cash income constraint for purchasing the consumption good (C), the farm production technology constraint, and the household total time endowment constraint. The last restriction (3.5) imposes non-negativity constraints on the labour categories, allowing households to optimally choose M, F or H to be zero.

It is analytically convenient to combine the three budget constraints represented by Equations 3.2 to 3.4 into a single constraint referred to as the *full income* constraint. This will be of the form [4]

$$(3.2n)\quad C + w\mathcal{L} = \{pQ - wF - w^hH\} + wT + E$$
$$= \Pi + wT + E$$

where Π represents the (short run) profit from farm production derived by deducting the cost of family and hired labour from gross output, with family labour valued at the market wage rate w. The total net returns to the farm household from its own farm production is broken down into two components: Π (which measures the net profit or returns from the ownership of the farm land), and wF (which represents the wage labour income from own-account work on the family farm).

The left-hand side of Equation 3.2n represents the total expenditure of the household on consumption of goods and leisure. The right-hand side represents the *full income* of the farm household; and it has three components: farm profit (Π), the value of the total time endowment of the household (wT) evaluated at the wage rate it receives when working off farm, and the non-labour endowment income, E.

To solve the household maximization problem specified above, substitute the constraint (3.4) directly into the utility function to set up the following Lagrangian:

(3.6)   Max $\pounds = U[C, (T - F - M)] +$
$$\lambda [pQ(F, H, A) - w^h H + wM + E - C] +$$
$$\mu_1 M + \mu_2 H + \mu_3 F$$

where μ_1, μ_2 and μ_3 are the Lagrangian multipliers associated with the non-negativity constraints on M, H and F, and which satisfy the following complementary slackness conditions:

(3.7a) $\mu_1 M = 0 \implies \mu_1 = 0$ if $M > 0$
(3.7b) $\mu_2 H = 0 \implies \mu_2 = 0$ if $H > 0$
(3.7c) $\mu_3 F = 0 \implies \mu_3 = 0$ if $F > 0$.

The choice variables for the household are C, M, F, H for which the first order conditions, respectively, are [5]

(3.8) $U_C = \lambda$

(3.9) $U_{\mathcal{L}} = \lambda w + \mu_1$

(3.10) $U_{\mathcal{L}} = \lambda (p \dfrac{\partial Q}{\partial F}) + \mu_3$

(3.11) $\lambda (p \dfrac{\partial Q}{\partial H} - w^h) + \mu_2 = 0$.

From (3.8) and (3.9) it follows that

(3.12) $\dfrac{U_{\mathcal{L}}}{U_C} = w + \dfrac{\mu_1}{\lambda}$.

The left-hand side of Eq. 3.12 represents the marginal rate of indifferent substitution of leisure for consumption ($MRS_{\mathcal{L}C}$) in the household's utility function. It measures the household's subjective valuation of leisure foregone in terms of the consumption numeraire when the household supplies an extra unit of labour. Let L represent the household's total labour supply ($L = F + M = T - \mathcal{L}$). Since,

(3.13) $\dfrac{U_{\mathcal{L}}}{U_C} = -\dfrac{U_L}{U_C}$

where U_L is the marginal dis-utility of labour, the left hand side of Eq. 3.12 measures the subjective value placed on a marginal unit of household labour. Sen (1966) refers to this as the "the real cost of family labour", whereas Nakajima (1986) calls it the "marginal (subjective) valuation of family labour". From Equations 3.8 and 3.10 it follows that

$$(3.14) \quad \frac{U_\ell}{U_C} = p\frac{\partial Q}{\partial F} + \frac{\mu_3}{\lambda} \quad ,$$

and Equation 3.11 can be re-written as

$$(3.11a) \quad p\frac{\partial Q}{\partial H} = w^h - \frac{\mu_2}{\lambda} \quad .$$

The farm household equilibrium with respect to labour supply and labour demand are then given by Equations 3.11a, 3.12 and 3.14.

These three equations represent the household labour supply and labour demand equilibrium for all possible combinations of the three labour categories (M, F and H). From Eq. 3.11a it is clear that if any hired labour is used (i.e. when H > 0, implying $\mu_2 = 0$), the value of the marginal product of hired labour is set equal to the wage rate paid out (w^h). And if no hired labour is used, it must be true that the value of the marginal product of the first unit of hired labour is less than w^h (since both λ and μ_2 are non-negative).

Similarly, from Equation 3.14, if no family labour supply is used for own farm cultivation (F = 0, implying $\mu_3 \geq 0$) then the household's valuation of its leisure must be no less than the marginal product for the first unit of family labour applied to the farm.

For the household that supplies some of its labour to own farm cultivation and some labour also to the off farm labour market at wage w, the household equilibrium is characterized by:

$$(3.15) \quad \frac{U_\ell}{U_C} = w = p\frac{\partial Q}{\partial F} \quad .$$

This is the standard first order condition which specifies that when family labour has two different uses – working on the family farm (F) and working in the off farm labour market (M) – the returns to labour in both activities are equalized at the margin. Family labour is applied in own farm production until the point where the value of the marginal product is equated to the market wage rate for off farm work.

For the optimal family labour supply not to involve any market work (i.e. when M = 0), the equilibrium condition is:

$$(3.16) \quad \frac{U_{\mathcal{L}}}{U_C} = w^* = p\frac{\partial Q}{\partial F}$$

where $w^* = w + \frac{\mu_1}{\lambda}$; and $w^* \geq w$, since $\lambda > 0$ and $\mu_1 \geq 0$ if M = 0.

The *w** represents a "virtual price" or "shadow wage rate" of family labour that applies to the subjective equilibrium of the labour-leisure choice faced by the farm household that only works on its own farm (Nakajima, 1986). If a farm household does not supply any family labour to the off farm labour market at a net wage of *w*, then the marginal returns to applying family labour to own farm cultivation must be no less than *w*.[6]

When M, F and H are all positive, such that all of the μ's are zero, the household equilibrium is characterized by the conventional optimality conditions. These equate the marginal products of the two types of labour inputs to their respective wage rates, and also the real cost of family labour to the market wage rate. These optimality conditions are given by Equation 3.15 and Equation 3.11a, with $\mu_2 = 0$.

Diagrammatic Illustrations

Figure 3.1 illustrates these equilibrium conditions for the allocation of family labour for the case where some family labour is applied on the farm and is also hired out (F > 0, M > 0). Figure 3.2 illustrates the case where all of the family labour supply is devoted to own farm cultivation (M = 0).

In Panel A of Figure 3.1 the vertical axis measures the value of consumption (C) in physical units (which is equivalent to a monetary unit, say Rupees, since p_c is normalized to one). The horizontal axis measures family labour input and the balance of leisure time, say in days. Since OT represents the total time endowment of family labour, the distance to the right from point O measures family labour workdays while the distance to the left from point T measures leisure days. The bold lines I_1 and I_2 represent the farm household's indifference curves between consumption and leisure. These indifference curves as drawn have the conventional property of being downward sloping and convex from below, given that in the {C, $\mathcal{L}$} space the origin is represented by point T.[7] The upward sloping curved line OGQ represents the farm production curve as family labour increases, measured in terms of the consumption good. Since p_c is set to one, OGQ represents the value of farm production as family labour input is increased, holding all other inputs (including hired labour) fixed.

The farm household equilibrium in Panel A is represented by the points of tangency of the line DGK with the farm production curve at the point G and with the indifference curve at the point J. The line DGK has the slope w, which is the off farm market wage rate for family labour. The tangency at point G represents the optimal family labour applied to own farm cultivation (in the amount of F) while the tangency at point J represents the total labour supply of the farm household (which includes F and the amount M of work in the off farm labour market at wage w).

Panel B of Figure 3.1 gives an equivalent representation of the labour-leisure equilibrium depicted in Panel A. The vertical axis of Panel B is in units of the C good while the horizontal axis is workdays, as in Panel A. The connection between the two Panels is that a measurement represented by a slope in Panel A becomes a vertical distance in Panel B. Hence, the YY curve in Panel B is the value of the marginal product of family labour which traces out the slope of the OGQ curve from Panel A at the equivalent level of family labour input.[8] Similarly, the VV curve measures the marginal rate of substitution between leisure and consumption which is the (positive) value of the slope of the indifference curve between leisure and consumption at different levels of family labour workdays. As drawn in Panel B of Figures 3.1 and 3.2, the VV curve is upward sloping in its entire range.[9] This need not be the case always; the VV curve can have flat sections, specially at low levels of labour supply. [10]

In Panel B the solution relating to the optimal use of family labour in own farm cultivation is given by point G' – the equality of the market wage rate and the marginal value product of family labour. (Point G' corresponds to the first equality in Eq. 3.15). The solution with respect to the optimal total labour supply of the farm household is given by point J' – the equality of the real wage rate (in units of the C good) and the marginal rate of substitution of leisure for consumption (which represents the real cost of family labour). (Point J' corresponds to the second equality in Eq. 3.15).

Panel B in Figure 3.1 also illustrates the recursive property of the farm household equilibrium when labour is supplied on the off farm hired labour market (i.e. when M > 0). The optimal allocation of family labour for own farm cultivation, represented by point G', is completely independent of the farm household's preferences with respect to leisure and consumption.

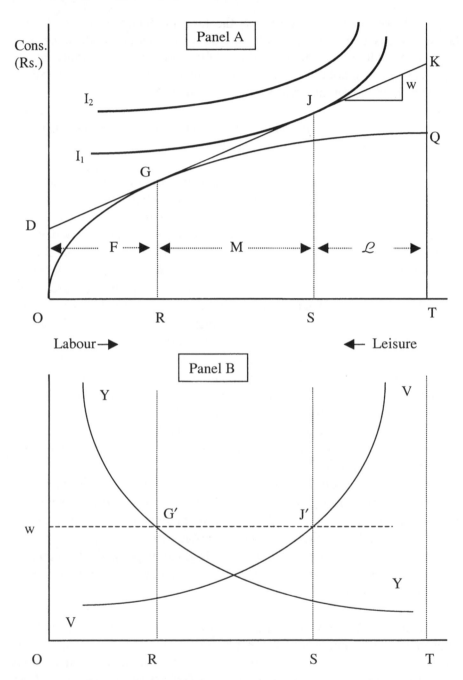

Figure 3.1 Farm Household Equilibrium with Labour Hired Out

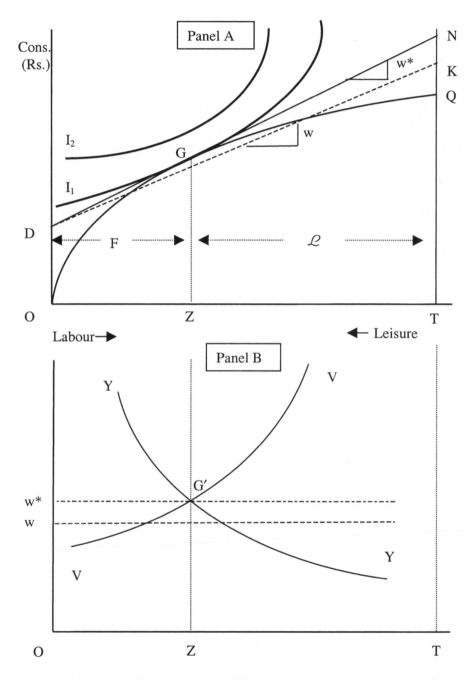

Figure 3.2   Farm Household Equilibrium with No Labour Hired Out

Such preferences, which affect the shape of the indifference curves, and hence the position and slope of the VV curve, determine only the total labour supply equilibrium position of J'. The VV curve has no effect on the position of G' which is determined solely by the given wage rate and the parameters of the farm production function which underlie the YY curve.

The equilibrium conditions for labour allocation of the farm household when all family labour supply is devoted to own farm cultivation (M = 0) is illustrated in Figure. 3.2. In Panel A the optimum allocation is given by the point G where the indifference curve I_1 is tangent to the value of farm production curve OGQ. The total labour supply of the farm household is OZ days and all of it is supplied to the own farm. At G the slope of the line of tangency DN, denoted as w^*, is higher than the slope of the wage line DK, given by w. When family labour is allocated to own farm cultivation only, the returns to own farm cultivation must be at least as great as the returns to working in the off farm labour market at the going wage rate (as indicated by the first order condition of Eq. 3.16 with $\mu_1 \geq 0$). The value of w^* can be interpreted as a shadow wage rate at which a hypothetical market for family labour is in equilibrium. A market wage rate of w^* would equate the demand for family labour in own farm cultivation to the supply of family labour offered at w^*. The equilibrium points G (and G' in Panel B) involve the farm household's subjective valuation of its labour which is determined jointly by production technology and preferences, and is independent of the observed market wage rate, w.

The farm household model with M = 0 is no longer separable into a producer and consumer equilibrium. The optimal input of labour on the family farm (which consists solely of family labour) is now affected by the household's preferences over leisure and consumption. Alternatively, the profit maximizing behaviour of the "farm-firm" cannot be separated out from the utility maximization behaviour of the consumer household.

3.3 Aggregating Labour Inputs and Production Function Separability

The farm household model presented in Section 3.2 above, with family and hired labour as completely distinct production inputs, is general enough to capture the various dimensions of any heterogeneity between them. As distinct inputs in the production function, no *a priori* restrictions are placed on how the marginal products of the two labour inputs are determined, nor on the elasticity of substitution between family and hired labour.

Nevertheless, there are several theoretical as well as practical difficulties in treating family and hired labour as completely distinct inputs. These problems are further compounded when family and hired labour can each be further dis-aggregated into several other meaningful sub-categories, e.g. male and female, or adult and child. While each of these labour sub-categories may not be perfectly homogeneous with all others, it is reasonable to expect many of them to be close substitutes for one another.

The wide diversity of labour sub-categories actually employed in peasant farming does not in itself mean each sub-category has a special role or a specific effect on production that cannot be matched by other sub-categories. Farm families will tend to use whatever family resources are available for work, especially with regard to child labour (White, 1994). This does not imply that child labour performs a special role that cannot be done as effectively by other adult family members.

Moreover, labour allocation in traditional agriculture reflects a wide variety of social and cultural norms which often give rise to specialized patterns of the division of labour among different labour categories, particularly by gender (K. Bardhan, 1993). For instance, in a given setting one specific farm operation may be done only by a particular labour category (e.g. weeding work done only by female labour while ploughing may be done only by males). Even if such an extreme division of labour occurs widely it is difficult to make the claim that any one sub-category of the many types of labour inputs observed is an *essential* input for farm cultivation.[11] In specific village settings, while there may be general conventions about the sexual division of labour and other labour categories, there often are sufficient exceptions to the rule to belie any claims for the *sui generis* nature of specific labour categories.

The main analytical problem with specifying a farm household model with many different labour categories is that the marginal product schedule for a specific type of labour (such as the YY curve in Fig. 3.1) can only be defined if the levels of the other sub-categories of labour inputs are taken as given. In an *n*-input production function, while the marginal product of a particular input is affected by the levels of all *n* inputs, the relationship is specifically sensitive to the levels of other inputs that are very close substitutes (or very strong complements). With several different labour categories that are close substitutes for each other and none of which is by itself an essential input, there is a wide variety of ways of generating the same level of an aggregate or composite labour input. The marginal product schedule for any one labour input category will depend not only on the substitution possibilities between that particular labour input and all other remaining labour categories, but also on the substitution possibilities among each of the remaining labour categories.[12]

While it is technically feasible to define a marginal product schedule for any one labour category, holding the levels of all other labour and non-labour inputs fixed, it is difficult to provide an economic explanation for why the other labour inputs are held constant at any particular level when there is a high degree of substitution among several labour categories. The marginal product schedule for family labour holding hired labour fixed at, say, 10 units will be very different from the one that holds hired labour fixed at 200 units. In order to make sense of the marginal product schedule for family labour around a specific equilibrium position it will be necessary to explain what is the optimal level of hired labour that goes with that particular equilibrium allocation of family labour. When family and hired labour are close substitutes for each other the relative allocation of the total labour input between family and hired labour may not be properly defined. Many different combinations of family and hired labour allocations could satisfy a given optimal level of the total labour input.[13]

A related problem is that with many distinct labour inputs, it is often difficult to rationalize the observed division of labour among various labour categories with the optimal conditions of the equality of the ratio of the marginal products with the ratio of the wage rates. The wage rates for different types of labour are often observed in fixed proportions. For instance, the female adult wage rate may be 80% of the male adult wage in a specific setting. However it would be difficult to justify that the underlying productivity differences between male and female labour should remain at a constant 80% in all activities undertaken by the farm household. In such a setting, a farm household model that had male and female labour as separate inputs would have difficulties in relating the observed division of labour by gender to the optimality conditions of the equality of the ratios of marginal products and the wage rates. With a fixed wage gap and close substitutability, one would expect such models to predict a much higher incidence of complete specialization – using only male or only female labour in farm cultivation – than what actually occurs.

There also are practical difficulties in estimating production functions with many categories of labour as distinct inputs. The parameter space expands rapidly with many distinct labour inputs, especially with flexible functional forms. There also are special econometric problems when many observations in a data set may have zero usage of some specific labour categories. A part of the empirical results later in Chapter 6 will show that the production function parameter estimates differ substantially between specifications that use distinct labour input categories and an aggregate composite labour category, even when using flexible functional forms.

It is meaningful then to seek to combine all the different labour sub-categories into a composite or effective labour input, while still allowing for heterogeneity between different labour categories. However, it is possible to create a consistent labour aggregate only under special separability properties of the underlying production function (Berndt and Christensen, 1974). Separability of the various labour input categories from other non-labour inputs in the production function implies that the marginal products of all the labour input categories are affected in an uniform manner by the levels of the non-labour inputs.

Formally, in an *n*-input production function given by

$$(3.17) \quad y = f(x_1, x_2, ..., x_n)$$

if the subset of labour input categories is indicated by the inputs x_1 to x_k the production function is "separable" in the labour sub-categories if the production function can be written in an equivalent form as

$$(3.18) \quad y = f\left\{ g(x_1, x_2, ..., x_k), x_{k+1}, ..., x_n \right\}$$

where the $g(.)$ function itself is quasi-concave and strictly monotonic.[14] The $g(.)$ function in itself has all the properties of a regular production function. It can be interpreted as a secondary level *micro* production function in which the output is the aggregate labour variable and the inputs are the different categories of labour.

Separability implies that the marginal rates of technical substitution between pairs of factors in the separated group are independent of the levels of factors outside that group.[15] Whether family and hired labour are separable from other inputs in a particular setting is a question that can be empirically verified with farm level data. If, in a given setting, the separability of the labour inputs can be established, then all the different categories of the labour inputs can be aggregated. This leads to considerable simplification in the analytical structure of the farm household model presented in Section 3.2 above, and consequently the separability of the production function greatly facilitates the subsequent empirical estimation even in the presence of labour heterogeneity.

The main theoretical implication of the separability of the labour inputs is that the ratios of the marginal product of family and hired labour (or alternatively, the marginal rate of technical substitution between family and hired labour) depend only on the levels of the labour inputs. Hence, optimal decisions about the allocation of family and hired labour can be made independently of the choices about other non-labour inputs. This gives rise to a sequential decision making process (Chambers, 1988: 45). With separability the optimal levels of the labour inputs are chosen only with respect to the relative price ratios (wage rates) of the different labour categories. Separability establishes a clear relationship between the shadow wage applicable to family labour and the market wage rate for hired labour on farms that use both family and hired labour. This relationship is derived formally in the next section.

3.4 A Farm Household Model with Heterogeneous Composite Labour

Assuming a separable structure of the labour inputs in the farm production function, the farm household's maximization problem can be set up in an identical manner to Equations 3.1 to 3.5 with the production function constraint (Equation 3.3) written in the separable format as

(3.3a) $Q = f\{g(F, H), A\}$

where $g(F, H) = Le$ defines an aggregate composite index of total labour input. Le can be interpreted as units of effective labour created out of the observed levels of family and hired labour applied in farm production.

The choice variables for the household utility maximization problem are the same as in Section 3.2; so the first order conditions are identical to Equations 3.8 to 3.11. With the production function being separable in the labour inputs, an additional property of the farm household equilibrium is that the ratio of the marginal product of family and hired labour is a function solely of the input levels of family and hired labour. That is,

(3.19) $\dfrac{\partial Q / \partial H}{\partial Q / \partial F} = \dfrac{\partial g / \partial H}{\partial g / \partial F} = h(F, H)$.

At the optimal values of family and hired labour input (indicated as F* and H*), let this ratio of the marginal products be given by θ*. That is,

(3.20) $h(F^*, H^*) = \theta^*$.

This additional condition derived from the separability of the labour inputs provides a way to relate the wage rates for the two types of labour, using the first order condition that the marginal product of hired labour will be set equal to w^h, the hired labour wage rate.

From the first order conditions for the farm household equilibrium, using Eq. 3.12 and Eq. 3.14, after substituting out for U_ℓ / U_C, gives

$$(3.21) \quad p \frac{\partial Q}{\partial F} = w + \frac{\mu_1}{\lambda} - \frac{\mu_3}{\lambda}; \quad \text{where } \mu_1 = 0 \text{ if } M > 0;$$
$$\mu_3 = 0 \text{ if } F > 0.$$

Equation 3.11a specifies that

$$(3.22) \quad p \frac{\partial Q}{\partial H} = w^h - \frac{\mu_2}{\lambda}; \quad \text{where } \mu_2 = 0 \text{ if } H > 0.$$

Hence, the ratio of the marginal products can be expressed as

$$(3.23) \quad \frac{\partial Q / \partial H}{\partial Q / \partial F} = \frac{w^h - \dfrac{\mu_2}{\lambda}}{w + \dfrac{\mu_1}{\lambda} - \dfrac{\mu_3}{\lambda}} = \theta^* = h(F^*, H^*)$$

keeping in mind the complementarity slackness conditions on μ_1, μ_2 and μ_3.

Equation 3.23 is the basic optimality condition that must be satisfied for all possible labour allocations involving any combination of F, M and H. Equation 3.23 is particularly useful in the case when family labour does not work at all on the off farm market at wage w (i.e. when M = 0) because in such situations the shadow wage rate for family labour can still be recovered from the observed levels of w^h. [16]

Consider first the case of a *big farm household* that must rely on hired labour to some extent (i.e. where $\mu_2 = 0$ because H > 0). For simplicity assume initially that $\theta^* = 1$ at the particular allocations of F and H that are observed. Then the shadow wage rate at the margin for valuing family labour, w^*, is given by

$$(3.24) \quad w^* = w + \frac{\mu_1}{\lambda} = \frac{w^h}{\theta^*} = w^h$$

Since one unit of family labour substitutes at the margin for one unit of hired labour which costs w^h, the relevant shadow price for family labour must also be w^h, as long as hired labour is employed on the farm and $w^h > w$. In this situation family labour is never supplied to the off farm market employment at wage w when it could be used to replace a unit of hired labour which costs w^h.

If instead $w > w^h$, then it would lead to complete specialization of the farm household labour allocation: all of its labour supply will be on market employment at wage w, and all of the labour input on the family farm will be hired labour paid a wage w^h per unit. Only in the case where $w = w^h$ would it be optimal for the big farm household to supply labour to the off farm hired labour market as well as on its own farm. This is readily evident from Equation 3.23 since for both F and M to be positive requires $\mu_1 = \mu_2 = 0$; hence if $\theta^* = 1$ it must be that $w = w^h$.

The farm household labour demand and labour supply equilibrium with $w^h > w$ (implying M = 0) and with $\theta^* = 1$ is illustrated in Figure 3.3. As in Fig. 3.1 and 3.2, the VV curve plots the rising real cost of family labour which is the utility cost of leisure foregone, represented by U_ℓ / U_C. The YY curve is the marginal product of aggregate labour drawn under the assumption that family and hired labour are homogeneous inputs ($\theta^* = 1$, everywhere). Since at the margin hired labour is used, the aggregate labour demand equilibrium is represented by the point P where the marginal product of the last unit of aggregate labour equals the hired wage rate. L_d measures the total labour days utilized on the farm, which under the assumption that $\theta^* = 1$ everywhere, is simply the sum of F and H in natural units. When using hired labour that is perfectly substitutable with family labour, the family labour supply equilibrium is represented by point E where $U_\ell / U_C = w^h$, and not point E_0 where $U_\ell / U_C = w$.

When $w^h > w$ and $\theta^* = 1$, the market wage (w) that family labour could earn when working off farm is *irrelevant* to the labour supply equilibrium of the big farm household that is hiring in labour at the higher wage rate, w^h. For every extra unit of family labour applied between E_0 and E, the real cost of labour is below w^h, while the gain in consumption is w^h – the saving on the wage payment for hired labour when family labour displaces it on a one to one basis. Consequently, the farm household attains a higher level of utility at point E than at E_0 (as long as E is a feasible point, with total workdays less than or equal to the total time endowment, T).

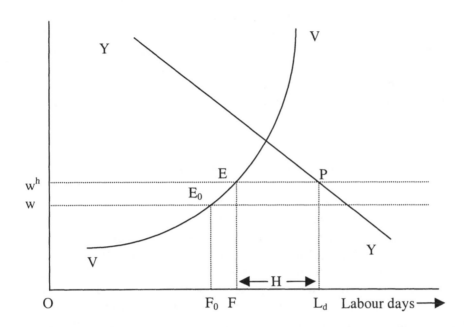

Figure 3.3 Labour Supply and Demand Equilibrium for a Big Farm Household

In the case of a *small farm household* the relevant family labour supply choice is to allocate the total time worked into the own farm labour input (F) component and the off farm market employment (M) component at wage w. In this situation also all the relevant scenarios can be derived from Equation 3.23. Again, when $\theta^* = 1$ everywhere, if $w > w^h$ one would observe complete specialization in the labour allocation of the small farm household also. All of its labour supply will be on market employment at wage w, and all of the labour input on the family farm will be hired labour paid a wage w^h per unit. Equation 3.21 is satisfied with $\mu_3 = 0$ and μ_1 and μ_2 non-negative. If $w < w^h$ then hired labour will not be used and family labour is allocated to the family farm till the marginal product of family labour is equated to the market wage rate. The remaining workdays are devoted to market wage employment. In either case the wage rate that applies to the determination of the household labour supply equilibrium is the market wage rate, w. The optimality condition derived from Equation 3.23 in this situation is exactly equivalent to the equilibrium conditions given by Equation 3.15 (derived in Section 3.2) where family and hired labour are treated as completely distinct inputs.

The more general case when $\theta*$ is different from one gives rise to a wide variety of alternative household labour supply and demand equilibria. These scenarios are discussed in detail in the next section where it is shown that the effective or shadow wage rate applicable for family labour supply decisions will depend on the hired labour market exposure of the farm household.

3.5 Labour Supply Implications of Heterogeneity

There are three different possible scenarios for the labour supply and demand equilibrium of the farm household when family and hired labour have intrinsic productivity differences ($\theta*$ is allowed to differ from one). These are summarized below. The main analytical point to note is that inherent differences in the productivity of family and hired labour are in essence equivalent to a wage gap for the services of hired and family labour (Benjamin, 1992). In the presence of labour heterogeneity, if at a particular labour allocation one unit of family labour would substitute for $1/\theta*$ units of hired labour, the implicit price the farm household would place on one unit of family labour would be $w^h/\theta*$, if hired labour was also being used by the household.

Case(i) : if $w > w^h/\theta*$

All farm households are completely specialized in their labour allocation. All days of family labour supply (L_S) are devoted to wage employment in the local labour market (at wage w); and all labour input (L_d) on the family farm consists only of hired labour (paid a wage w^h). In this scenario it is not meaningful to distinguish between small farm households who are net sellers of labour and big farm households who are net buyers of labour. The optimal labour supply and demand conditions for all households, irrespective of their total labour demand, are identical and are given by

$$(3.25) \quad U_\ell/U_C = w \; ; \qquad L_S = M , \; (F = 0) ;$$

$$(3.26) \quad p\frac{\partial Q}{\partial H} = w^h \; ; \qquad L_d = H .$$

These optimal conditions are independent of the value taken by θ^*. When $w > w^h/\theta^*$, irrespective of the value of θ^*, it pays the farm household to specialize by offering all of its desired labour supply on the off farm labour market, while hiring in all its labour demand requirements.

In this scenario, in spite of the presence of labour heterogeneity ($\theta^* \neq 1$), the consumption and labour supply component of the farm household's decision is independent of the production side choices for all households. The effective wage rate that determines the farm household's optimal labour/leisure choice is the market wage rate, w; and the farm household model is recursive.

Case(ii) : if $w = w^h/\theta^*$

When the differences in the wage rates for family and hired labour exactly offset the difference in productivity, the equilibrium conditions are :

$$(3.27) \quad w^h = p\frac{\partial Q}{\partial H}$$

$$(3.28) \quad U_\ell/U_C = w = \frac{w^h}{\ulcorner^*} = p\frac{\partial Q}{\partial F} \quad .$$

When any observed efficiency difference between family and hired labour is exactly mirrored in the difference in the market wage rates available to hired and family labour, the family and hired labour inputs become perfect substitutes (in the neighbourhood where $w = w^h/\theta^*$). This sub-case reduces to the conventional farm household model where family and hired labour are treated as homogenous inputs with a common market wage rate. The only adjustment is in the units of measurement – i.e. whether total labour is measured in units of family or hired labour days, with θ^* being the conversion factor for the quantities and the wage rates.

In this scenario, as in the case of homogeneous labour inputs, the farm household model cannot be solved for specific levels of F, H or M. What is determined is total labour demand in effective units (F + θ^*H); but its composition between F and H is indeterminate. Similarly on the consumption/leisure component of the model, only total labour supply days (F + M) is determined; its allocation between F and M is arbitrary. In this situation altering the mix between F and H and between F and M, while keeping total labour supply and demand fixed, has no effect on farm profits or wage income, and hence no effect on the household utility level.

Case(iii) : if $w < w^h/\theta*$

In this case the analytical structure of the farm household differs according to the net position of the household in the hired labour market. Three distinct farm household types can be distinguished, and the effective wage rate that determines the labour supply equilibrium will differ for each.

(a) <u>the small farm household</u> whose total labour supply will consist of work on the family farm (F) supplemented by off farm market wage employment (M) at wage w. The effective wage rate for this household is the market wage rate, w; and its equilibrium position is given by

$$(3.29) \quad U_{\ell}/U_C = w = p\frac{\partial Q}{\partial F} \quad ; \quad M > 0; F > 0; H = 0 \, .$$

These conditions are exactly equivalent to the equilibrium conditions for a household with $M > 0$ given by Equation 3.15 in Section 3.2, where family and hired labour are treated as completely distinct inputs. From Equation 3.29, this sub-model is also clearly recursive. No hired labour is used because the total family labour supply exceeds the total labour demand and the wage cost of family labour is lower than that of an equivalent unit of hired labour. It will be more profitable for the farm household to use family labour valued at wage w than hired labour at an effective wage of w^h/θ.

(b) <u>the autarchic household</u> which neither sells any family labour on the market nor employs any hired labour (i.e. $M = H = 0$). Its equilibrium is described by

$$(3.30) \quad \frac{U_{\ell}}{U_C} = w* = p\frac{\partial Q}{\partial F}$$

$$(3.31) \quad \text{where} \quad w \leq w* \leq \frac{w^h}{\epsilon *} \, .$$

The effective wage rate, $w*$, faced by this household is indeterminate within the bounds specified above.[17] Since $w*$ is partly a function of the marginal product of family labour which depends on production technology, the autarchic farm household model is not recursive.

(c) the big farm household which must supplement its family labour days with hired labour in order to meet the total demand for labour in its farm cultivation. Given that $w < w^h/\theta*$, the big farm household never supplies any labour to the off farm market at wage w since that same unit of labour when applied to the family farm earns a return of $w^h/\theta*$ when family labour can substitute for $1/\theta*$ units of hired labour.

The optimum labour input allocations will satisfy

$$(3.32) \quad w^h = p\frac{\partial Q}{\partial H}$$

$$(3.33) \quad U_\ell/U_C = w* = p\frac{\partial Q}{\partial F} \quad ; \quad F > 0; M = 0 ; H > 0;$$

where $w \leq w* = \dfrac{w^h}{\Gamma*}$, since by assumption

$$(3.34) \quad \frac{w^h}{w*} = \left(\frac{\partial Q/\partial H}{\partial Q/\partial F}\right)_{F*H*} = \theta*.$$

Consequently, for a household which employs some hired labour the shadow wage rate $w*$ which determines its own labour supply equilibrium is not "subjective" any more as it was in the general model with family and hired labour as distinct inputs.[18] With the production function being separable in the labour inputs, this shadow wage rate can be related directly to the wage rate for hired labour with the adjustment for the difference in the marginal productivity, if any, of the two types of labour at that particular equilibrium. Hence, $w* = w^h/\theta*$.

The farm household sub-model for the big farm household that hires in labour is not generally recursive since the effective wage rate for family labour, $w^h/\theta*$, is not exogenously given to the household. It will depend on the allocation of F and H which affects $\theta*$. If however, within the relevant range of labour allocations for a particular household, $\theta*$ happens to be constant then the sub-model for the big farm household becomes recursive. In this situation the household equilibrium will be identical to the case of a household facing an exogenously given market wage rate of w^h for hired labour and a market wage rate $w^h/\theta*$ for family labour.

Diagrammatic Illustration for Case (iii) when $w < w^h/\theta*$

The labour supply and demand equilibrium for these three types of households are illustrated in Figure 3.4 for a specific scenario where $\theta* < 1$ and $w < w^h < w^h/\theta*$. The marginal product of family labour is denoted by the Y_F schedule while the marginal product of hired labour is denoted by the Y_H schedule. As drawn Y_H is always below Y_F indicating that at any given labour allocation of F and H, adding one more unit of F leads to a higher increase in output than adding an extra unit of H. This implies $\theta* < 1$ in the entire range of total labour demand (which, in what follows, is denoted simply as the condition $\theta < 1$, for all combinations of F and H).

In each panel the point labelled E denotes the appropriate labour supply equilibrium. Hence, in estimating the labour supply component of the farm household model the wage rate variable used in the regression equation must conform to the wage rate relevant for point E. This wage rate varies for the three different household types[19] as indicated in Panels (a), (b) and (c) of Figure 3.4, respectively for the small farm, autarchic and big farm households. The effective wage rate for family labour is higher than the market wage rate in the autarchic and big farm households.

Panel (a) shows that the differences in the productivity of family and hired labour have no effect in the neighbourhood of the labour supply equilibrium of the small farm household. At the margin it faces a given wage rate which determines the total amount of its labour supply. The Point E in Panel (a) is unaffected by the value of θ. The only relevance of productivity differences in Panel (a) is that with $\theta < 1$ and $w < w^h/\theta$, the small farm household does not simultaneously use hired labour as well as supply its own labour to off farm employment.

In Panel (b) the farm household is autarchic: it neither buys any hired labour nor does it sell any of its family labour on the off farm market. Total household labour supply exactly matches the total demand for labour in farm production (the amount OF). The household acts as if it were facing an effective wage rate $w*$ which equates the demand and supply of family labour. This household does not sell any of its family labour on the market because the returns to labour on own farm cultivation are higher than the market wage rate. On the other hand, given the family labour input of OF, the marginal product of an extra unit of hired labour is less than the hired labour wage rate w^h.

As drawn in Panel (b) the autarchic equilibrium is created by the fact that $\theta < 1$. The wage gap between w and w^h by itself is insufficient to create an optimal autarchic equilibrium since $w* > w^h$ as drawn. If $\theta = 1$ and the marginal productivity of both family and hired labour were to be represented by the Y_F schedule, it would not be optimal for this household to be autarchic.[20] In that case family labour supply to own farm cultivation would only be OF^0 (instead of OF) and a few extra units of labour would be hired in (up to the point of intersection of the Y_F schedule and the w^h wage line).

Finally in Panel (c), which represents the big farm household, the effective wage rate applied to family labour at point E is directly affected by the value of θ. The fact that $\theta < 1$ has two separate effects on the optimal labour supply and demand choices of the big farm household:

i.   the first effect is that total labour demand is reduced (from OL_d^0 to OL_d). Without any differences in the productivity of the two types of labour, the Y_H schedule would coincide with the Y_F schedule and the optimal total labour demand would be the point of intersection of the labour marginal productivity schedule with the w^h wage line (which gives total labour demand of OL_d^0).

ii. the second effect is a higher labour supply response for family labour because it leads to a higher effective wage rate for family labour. Family labour supply is OF with $\theta < 1$ as compared to OF^0 with $\theta = 1$.

The combined effect of a lower total labour demand and a higher proportion of it coming from family sources means that the demand for hired labour is considerably reduced on big farms because of the productivity difference. The amount of hired labour used on the big farm household illustrated in Panel (c) is the distance between F and L_d (indicated as H). If family and hired labour were homogeneous inputs the demand for hired labour on this big farm would instead be equal to the distance between F^0 and L_d^0.

As discussed in Chapter 2, one of the key stylized facts about agricultural production in developing countries is that small farms are cultivated more intensively with higher levels of variable inputs used per hectare than bigger farms. This often leads to the celebrated observation of an inverse relationship between farm size and output per hectare. The relatively greater application of inputs on smaller farms is most pronounced in the case of labour. The per hectare labour input on small farms is consistently higher than on bigger farms over a large range of farm sizes; and this result holds whether or not average yields on small farms are

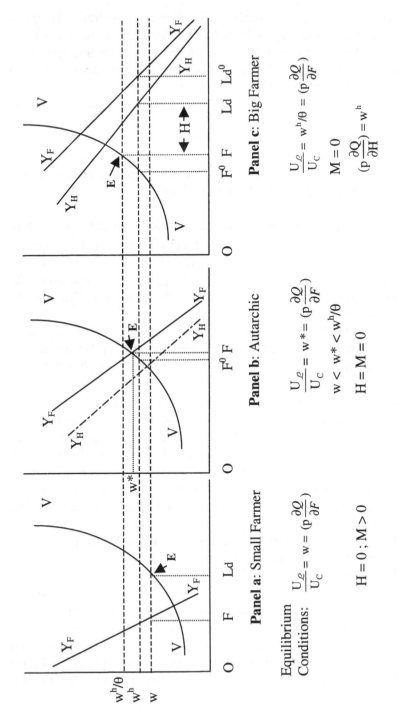

Figure 3.4 Farm Household Equilibrium with Labour Heterogeneity

higher (Berry and Cline, 1979). A traditional explanation for the higher labour intensity on small farms has focussed on factor market imperfections (Sen 1966, 1975). By comparing Panels (a) and (c) of Figure 3.4, it is clear that efficiency differences between family and hired labour can be a source of the inverse relationship even if rural labour markets are functioning well and the market wage rates are set in a competitive manner.

The equilibrium of the three types of farm households illustrated in Figure 3.4 for the Case (iii) scenario of $w < w^h/\theta$ suggests that efficiency differences in the productivity of family and hired labour are analytically equivalent to a lower market wage rate for family labour than for hired labour. The three types of households in Figure 3.4 can still be identified even if $\theta = 1$ as long as $w < w^h$, but the equilibrium will be at a different position in Panels (b) and (c). The equilibrium position is unaffected by the θ parameter only in the case of the small farm household of Panel (a).[21]

Such an equivalence between an observed market wage gap for family and hired labour and inherent productivity differences suggests that the labour supply component of the farm household model will not be able to distinguish between these alternative explanations for the higher effective wage rate for family labour in autarchic and big farm households. If there is any labour heterogeneity that leads to efficiency differences between family and hired labour ($\Box^* \neq 1$), it must be detected directly from the production function estimations. If in addition a wage gap is observed between w and w^h, this additional information must be taken into account in deriving the effective wage rates to describe the labour supply equilibrium positions.

3.6 Summary

This chapter presented the analytical framework for a farm household model allowing for family and hired labour to be heterogeneous inputs. A farm household model based on a general form of labour heterogeneity that allows for family and hired labour (as well as other relevant labour categories) to be completely distinct inputs is analytically unappealing and empirically difficult to estimate. A particular problem in treating family labour as a distinct input is that the wage rate for determining the labour/leisure equilibrium choice of the household is unidentified when family labour is devoted solely to own farm cultivation.

An alternative approach to modelling labour heterogeneity within a composite effective labour input is more appealing. With such an approach the effective wage rates for family labour can be related directly to the observed market wage rates at which non-family labour can be hired in. A composite labour input is defined if individual labour inputs are separable from the other inputs in the production function. If this condition is met the ratio of the marginal products of family and hired labour becomes independent of the levels of the other non-labour inputs. Consequently, the first order conditions for the optimal use of hired labour can be used to derive the effective wage rates for family labour as well. This framework is general enough to accommodate wage gaps that arise when the hiring in wage for non-family labour and hiring out wage for family labour differ.

The separability of the labour inputs in the production function in essence makes family labour equivalent to a traded input even when family labour is supplied only to own farm production work. Such an equivalence is obtained by comparing the marginal product of family labour with that of the actually traded input (hired labour). This relationship can be used to derive the effective wage rate for family labour on large farms that also employ hired labour. Such a procedure considerably simplifies the estimation of the labour supply component of the farm household model for households engaged in own farm work only.

The predictions made by a model of the farm household with heterogeneous labour inputs are testable. When $w > w^h/\theta*$, it should lead to complete specialization in the labour supply and demand of all farm households. This situation, however, is rarely observed in practice. In the more likely scenario of $w < w^h/\theta*$, it gives rise to three distinct types of farm households: the small farm household that does not employ any hired labour, but sells surplus family labour on the off farm market; an autarchic household which neither buys nor sells any labour; and a big farm household which devotes all of its family labour to own farm cultivation, supplemented by hired labour. The effective wage rate that identifies the labour/leisure equilibrium for all three household types can be identified, and an estimation strategy devised based on observed market wage rates and estimated parameters of the production function.

Notes

1   Becker's (1965) new household economics is a separate development from farm household models that has been applied mainly to developed country settings.

2   See Chiappori (1988, 1997) for household models with individual bargaining.

3   The main analytical relevance of explicitly modelling other variable inputs is that the presence of other variable inputs may affect the marginal product of the two types of labour in different ways. If the ratio of the marginal product of family labour to that of hired labour is sensitive to the level of other inputs, the production function is not *separable* in the two labour inputs (Chambers, 1986: 43). This means the two labour inputs cannot be consistently aggregated into a composite labour input. Whether family and hired labour are separable inputs in a particular setting is an empirical question; and this is addressed in Chapter 6 for the data set from the *tarai* region of Nepal. Note this use of the term "separable" as a property of the production function is different from the "separable" (recursive) property of farm household models.

4   Equation 3.2n is derived by adding $w\mathcal{L}$ to both sides of Equation 3.2 and substituting out for $\mathcal{L}$ on the right hand side using Equation 3.4.

5   Following conventional notation $U_C = \partial U / \partial C$ and $U_\mathcal{L} = \partial U / \partial \mathcal{L}$.

6   This result holds under the assumption there are no quantity constraints on the days of labour supplied and no fixed costs to working on the off farm labour market.

7   Along the indifference curves I_1 and I_2, as leisure increases the slope of the indifference curve is falling, indicating a decreasing marginal rate of substitution between consumption and leisure as more leisure is taken. More formally, the sufficient conditions to ensure that indifference curves are downward sloping in the $\{C, \mathcal{L}\}$ space and convex to the origin are

(N3.7.1) $\dfrac{\partial U}{\partial C} = U_C > 0$; $\dfrac{\partial U}{\partial \mathcal{L}} = U_\mathcal{L} > 0$;

(N3.7.2) $\dfrac{\partial (U_\mathcal{L}/U_C)}{\partial C} = > 0$; $\dfrac{\partial (U_\mathcal{L}/U_C)}{\partial \mathcal{L}} = < 0$.

The implications of these restrictions are fully drawn out in Nakajima (1986: 11-14). Although Nakajima works directly with a utility function specified in terms of C and L (labour workdays) instead of C and $\mathcal{L}$ (leisure days) as in the above, the restrictions above are exactly equivalent to restrictions 2.2, 2.11 and 2.12 imposed by Nakajima to obtain well behaved indifference curves which slope upwards in the $\{C, L\}$ space and which are convex from below.

8   The YY curve is downwards sloping in its entire range under the assumption of a regular production function with declining marginal productivity of all inputs, including family labour.

9   The VV curve sloping upward in its entire range implies that the $MRS_{\mathcal{L}C}$ is continuously increasing as workdays (L) increases or leisure ($\mathcal{L}$) decreases. Following Nakajima (1986: 26) the necessary and sufficient conditions for VV to be upward sloping are :

(N3.9.1) $Z = \dfrac{U_\mathcal{L}}{U_C} > 0$; and $\dfrac{\partial Z}{\partial C} * p \dfrac{\partial Q}{\partial F} - \dfrac{\partial Z}{\partial \mathcal{L}} > 0$.

These conditions are satisfied under the restrictions on the utility function specified by N3.7.1 and N3.7.2 in Endnote number 7 above (assuming that the marginal product of family labour, $\partial Q / \partial F$, is always non-negative).

10   It is quite feasible that VV is horizontal at low levels of labour and consumption. For instance, if one postulates a minimum subsistence consumption level C_0, and given that the low levels of labour input will lead to low levels of C (ignoring non-labour income), the MRS_{LC} may remain unchanged until the labour days worked is sufficient to attain the C_0 level of consumption. Graphically this means the indifference curves I_1 and I_2 when transferred to the region below C_0 would be upward sloping straight parallel lines with a constant slope in the {C, L} space. Consequently in Panel B, there would be a section of the VV curve being horizontal at low levels of labour supply. See Nakajima (1986: 19-20). Horizontal sections of the VV curve are of analytical interest since it gives rise to "surplus labour" in the sense defined by Sen (1966).

11   An essential input must be used in order to have positive levels of output. If $F(x_1, x_2, ..., x_n) = 0$ whenever $x_i = 0$, then x_i is an essential input (Chambers 1988: 9).

12   In an *n*-input production function with *k* labour categories, there are *k* times $(k-1)/2$ distinct partial elasticities of substitution among the labour inputs themselves. If even only a few of these distinct partial elasticities are high, it becomes difficult to derive a marginal product schedule based on only one specific labour category without aggregating in some way all the other remaining categories.

13   In the conventional farm household model where family and hired labour are treated as homogeneous inputs, the model determines only the total labour demand and labour supply. Any combination of family and hired labour on the demand side and any matching combination of own farm work and off farm work on the labour supply side can satisfy these aggregate levels. Such indeterminacy of the individual components of F, H and M arise as the degree of substitution between family and hired labour increases even when family and hired labour are treated as distinct inputs.

14   Goldman and Uzawa (1964).

15   The formal representations of the separable property of a production function are discussed in more detail in Chapter 6. Note that "separable" as a property of the production function is different from the "separable" (recursive) property of farm household models. To avoid confusion henceforth, the term "recursive" will be used to refer to the property of farm household models while the term "separable" will refer only to the property of production functions.

16   The separability of the labour inputs in the production function is not a necessary condition to relate the shadow wage rate of family labour to the wage rate for hired labour. However, without the separability property the ratio of the marginal products of family and hired labour are functions of the level of other non-labour inputs as well; so this relationship will be less tractable.
      Under non-separability of the production function given by Equation 3.3, the specification for the ratio of marginal products of the two labour inputs becomes $\theta^* = h(F^*, G^*, A^*)$ where A is the land input. Although the shadow wage rate for family labour (w^*) can still be related to w^h and θ^*, the value of w^* will vary among farmers who hire in labour at the same wage rate but whose farm size (A) differs. When the production function is separable, θ^* will be independent of A and all other non-labour inputs. Hence w^* can be related directly to w^h and the levels of F and H only.

17   The bounds on w^* are slightly wider than is specified by Equation 3.31 when $\theta^* > 1$. If hired labour has a higher marginal product the farm household can be autarchic only if there is no net gain from transferring one unit of family labour from own farm work to market work, while replacing the unit of family labour with hired labour for farm cultivation. This leaves unchanged the total labour supply of the farm household.

Hence the net gain can be determined by the change in consumption which is given by:

$$(N3.17a) \quad (w - p\frac{\partial Q}{\partial F}) + (p\frac{\partial Q}{\partial H} - w^h) \ .$$

A farm household can remain autarchic only if the above expression is non-positive.

Noting that $w^* = p\frac{\partial Q}{\partial F}$ and $p\frac{\partial Q}{\partial H} = \Box^* p\frac{\partial Q}{\partial F}$, the required necessary condition for an autarchic state to be optimal reduces to

$$(N3.17b) \quad w \leq w^* \leq \frac{w^h - w}{\Box^* - 1} \quad \text{if } \theta^* > 1.$$

It can be verified that $\dfrac{w^h - w}{\Box^* - 1} > \dfrac{w^h}{\Box^*}$ if, as assumed, $\theta^* > 1$ and $w < w^h/\theta^*$.

18 From Equation 3.16 $w^* = w + \dfrac{\mu_1}{\lambda}$ when family labour is modelled as a distinct input and all family labour is supplied on the family farm. Since μ_1 and λ are Lagrange multipliers that vary with the constraints of the model, the effective wage rate w^* is subjective and cannot be identified from the observed market wage rate w.

19 Figure 3.4 is drawn with $w < w^h$. This is not a necessary condition for the three different household types to be defined. The necessary condition is $w < w^h/\theta$. When $\theta < 1$, this condition can be satisfied even if $w > w^h$.

20 With $\theta = 1$ an autarchic equilibrium could still be optimal if the intersection of the Y_F and V schedules was such that $w < w^* < w^h$.

21 This ignores potential general equilibrium effects whereby the demand for hired labour is affected by the θ parameter, and hence has an effect on the equilibrium market wage rate at which a small farm household can work off farm.

4 Estimation Strategy

This chapter develops the econometric strategy for both detecting and incorporating potential labour heterogeneity in the production and labour supply component of the farm household model. The model estimation results, based on household survey data from the *tarai* region of Nepal, are presented in Chapters 6 and 7. In this chapter, Section 4.1 reviews the general estimation issues involved and briefly discusses alternative strategies. Section 4.2 outlines the sequential estimation strategy proposed by Jacoby (1993) for estimating non-recursive farm household models. Section 4.3 presents an adaptation of this methodology for estimating a farm household model with labour heterogeneity that is consistent with the structure of a farm household model with heterogeneous composite labour developed in Chapter 3. Section 4.4 contains a brief discussion of the required adjustments to the variance-covariance matrix of the parameters estimated in the second step of a sequential estimation strategy.

4.1 General Issues

The conventional approach to estimating farm household models treats family and hired labour as homogenous inputs valued at a common wage rate. Consequently a two step estimation procedure is feasible. In the first step, the production side of the model is estimated either with a production function (Barnum and Squire, 1979), a profit function (Sidhu and Baanante, 1981), or a cost function (Binswanger, 1974). These functions are specified for a given level of commodity dis-aggregation for outputs and inputs. The labour input variable is total labour: its composition between family and hired labour need not be addressed. The first step produces a set of output supply and input demand equations as functions of input and output prices, fixed inputs and other farm characteristics.

In the second step, a consumer demand system is estimated to represent the farm household's preferences. This second component can be modelled as a complete dis-aggregated demand system (such as the Linear Expenditure or Log Linear Expenditure system) that specifies leisure as one of the consumption goods (Kuroda and Yotopoulos, 1980). An alternative approach is to directly specify a reduced form labour supply equation (Rosenzweig, 1980) which can be made consistent with a demand system in which leisure is separable from other consumption goods.

Irrespective of the approach and specific functional forms chosen for the production and consumption components of the model, the distinctive feature of the conventional estimation strategy is that the consumption/labour supply choices are estimated independently of the production side of the model, relying on the recursive property of the standard farm household model. The only connection between the production and consumption side in the conventional estimation of the recursive model is that the consumption choices are conditioned on the level of "full income" of the household, which includes the value of farm profit at the production optimum.[1] On the consumption side, the commodity demands and labour supply are functions solely of commodity prices, including wage rates for labour, household full income, and possibly some household and individual characteristics that affect preferences. Hence, the full consumer demand system or the labour supply equation alone can be estimated in an independent fashion based solely on exogenous variables and the farm profit variable which can be computed from the observed production side choices (Singh, Squire, and Strauss, 1986c1: 20).

On the other hand, when family and hired labour must be treated as heterogeneous inputs, in the general case the recursive structure of production and consumption choices breaks down.[2] The econometric specification of fully non-separable models is analytically cumbersome and difficult and time intensive to empirically implement. Joint estimation of farm production, input demands as well as consumer demand equations may not even be feasible at times. It may be impossible to solve for the reduced forms of these equations analytically (Singh, Squire and Strauss, 1986c1: 22). Given that a non-recursive farm household model would be highly non-linear in the parameters, it may not be feasible to identify the structural parameters of interest even when it is possible to solve for their reduced forms (Jacoby, 1993). A joint systems estimation relying on full information methods will be more cumbersome as the level of commodity dis-aggregation increases. Moreover, the data requirements for the joint estimation approach are enormous. Data are required for all endogenous variables in the system (in addition to the set of properly exogenous variables); while in a recursive model estimation can be done with limited data sets that cover only specific components of the farm household model.

While non-recursive models are difficult to implement empirically, there are costs to assuming a recursive estimation strategy when the underlying model is genuinely non-recursive. The interdependence of the production and consumption choices of farm household models affects empirical estimation based on an assumed recursive structure in two ways.[3] The usual parameters estimated in the output supply, factor demand and consumer demand systems that incorrectly assume a recursive structure are statistically inconsistent. Secondly, these parameters by themselves are unable to identify the full effect of the comparative statics of the truly non-recursive model because the latter model will have additional terms that have not been estimated.[4] The total error resulting from the combination of the inconsistent estimates and the missing terms in the comparative static derivations will be difficult to gauge.

Previous studies have not adequately looked into the question of how large the errors are in implementing a recursive estimation strategy when the underlying model is truly non-recursive. Lopez (1984) – which is also summarized in Lopez (1986) – is the only example where the results of a full information joint estimation strategy has been compared with the results from assuming the conventional recursive farm household model structure. These show significant differences in the values of the parameters of the farm household model when using a recursive *versus* a non-recursive estimation strategy. For instance, the estimates of the own wage elasticity of family labour supply changes from 0.19 in a recursive estimation strategy to 0.04 in a non-recursive strategy. While the differences in estimates are numerically large, Lopez unfortunately does not report standard errors for each set of estimates. So one cannot test whether the differences are statistically significant. A second important issue which was also not addressed is whether the differences in these two estimates are also of economic significance – i.e. do they vary enough to lead to significantly different results on specific applications of the model, especially on implied policy prescriptions. Again, there is little empirical evidence on how significant such model estimation errors can be in a substantive sense.

4.2 The Two Step Estimation Strategy for a Non-Recursive Model

The possibility that estimation errors in assumed recursive models could be significant – statistically and substantively – but not always large enough to warrant full-scale joint estimation has led to the search for alternative estimation strategies. The objective has been to develop procedures that are valid for a non-recursive specification but which retain the tractability of the stepwise estimation of the production and consumption/labour supply components of the conventional farm household model.

One such approach has been proposed and applied by Jacoby (1993). Jacoby's approach is a general methodology for estimating a structural labour supply equation for workers who are self-employed. It is analogous to the treatment of labour supply in the presence of progressive income taxes that was pioneered by Hall (1973) which involves "linearizing" an underlying non-linear budget constraint. This approach is widely used in labour supply estimation in developed country settings,[5] but Jacoby's application to a farm household setting is novel.

In the context of farm households whose members are purely self-employed ($M = 0$ in the terminology of Chapter 3), the household budget constraint for consumption is given by net farm output, which is an increasing function of family labour days applied to farm production. Consequently, the household faces a concave budget constraint in the form of a well behaved agricultural production function with diminishing marginal product of family labour. The methodology proposed by Jacoby is to linearize the budget constraint at the household's optimum equilibrium point. The objective is to convert the farm household's labour/leisure choice problem with a concave budget constraint into an equivalent standard consumer problem where the farm household acts as if it faced a linear budget constraint with a fixed wage rate which makes it choose the same optimal labour supply position as with the concave budget constraint. Figure 4.1 illustrates this linearization process.

For the simple farm household model presented in Chapter 3 (where labour and land are the only production inputs), from Equation 3.2, the household's budget constraint for consumption is given by

(4.1) $C = (pQ - w^h H) + wM + E$.

The three components of the budget constraint represented above are the net returns from own farm cultivation ($pQ - w^h H$), wage income when family labour works on the off farm labour market (wM), and non-labour endowment income (E) which is exogenously given. The full range of such a budget constraint for the farm household is diagrammatically represented in Figure 4.1 by the solid bold line OBGPK.

The OBGPK line has three segments to match the three components of the household's budget constraint in Equation 4.1. The vertical segment OB measures the level of consumption that can be funded solely from the household's non-labour endowment income, hence the height of OB measures the value of the variable E in Equation 4.1. The curved segment BGP represents net income from farm production.[6] It includes the returns to the ownership of the land as well as to family labour applied to own farm production. As drawn, BGP is a part of the farm production function for increasing levels of family labour application (holding all other inputs constant) which is represented in its entirety by BGPQ. Hence, the slope of BGP is the marginal product of family labour which is decreasing as more family labour is allocated to farm production. The final linear segment PK represents labour income from off farm market work. The slope of PK is the exogenously given market wage rate, w.

Given the household budget constraint line OBGPK and household preferences over leisure and consumption denoted by indifference curves such as I_1, (and assuming that at least some family labour is devoted to own farm production),[7] two equilibrium positions are possible. The tangency of the indifference curve can be either with the curved section BP of the budget constraint or on the linear section PK.

The scenario that is drawn in Figure 4.1 is of a farm household equilibrium on the curved section BP of the budget constraint. Given household preferences, the production technology and the market wage rate (w), the household finds it optimal to supply all its labour to own farm production and not engage in any off farm work. This optimal labour allocation is denoted by the point G. At this equilibrium the subjective valuation of family labour (U_ℓ / U_C) is set equal to the marginal product of family labour. The latter is given by w^*, the slope of the farm production function at G. Hence the optimal allocation of family labour at point G is supported by a shadow wage rate of w^*, that is necessarily greater than or equal to the market wage rate, w.[8]

The process of linearizing the household budget constraint at the optimal point G in Figure 4.1 proceeds as follows. If the concave budget constraint OBGPK that the farm household actually faces were to be replaced by an artificial linear budget constraint represented by the dashed bold line DGN in Figure 4.1, the farm household equilibrium remains unchanged at point G. This artificial linear budget constraint DGN is uniquely defined by two parameters: (i) its slope w^*; and (ii) the intercept OD. The intercept OD represents the consumption level that could be attained, based on the fictitious linear budget constraint line DN, if all available family labour time was devoted to leisure.

Hence, if the budget constraint of this particular farm household is re-formulated in a linear form, uniquely defined by the slope w^* and intercept OD, the equilibrium point G would represent the optimum solution to the standard consumption/leisure choice problem faced by a household that was not involved in its own farm production. G represents the labour supply equilibrium of a household with a fixed labour endowment (T) and an exogenously given non-labour endowment income (equal to the amount OD), where the household faces an exogenously given market wage rate of w^*, which it takes as given for all levels of its labour supply.[9]

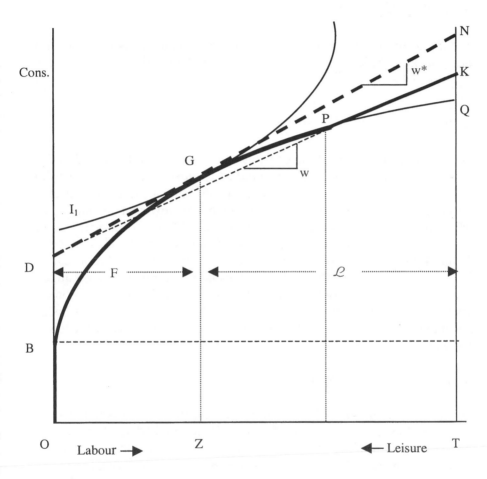

Figure 4.1 Linearized Budget Constraint for a Farm Household Equilibrium

That is, the point G is also the solution to the problem

(4.2)   max U$(C, \mathcal{L})$
subject to

(4.3) $C = w^*(T - \mathcal{L}) + E^*$

for which the first order condition

(4.4a) $U_{\mathcal{L}}/U_C = w^*$

is identical to the first order condition for the labour supply component of the farm household's utility maximization problem (the first equality in Equation 3.16) where [10]

(4.4b) $w^* = w + \dfrac{\mu_1}{\lambda}$.

This re-formulation of the consumer problem with a linearized budget constraint in essence *takes the farm out of the farm household's decision-making locus*. The farm household becomes equivalent to the standard consumer household with a given non-labour endowment income (OD), and a labour time endowment (T) that can be sold at a fixed market wage rate, w^*. This is the conventional household of labour economics textbooks (Killingsworth, 1983) for which an internal equilibrium solution can be derived in a much more straightforward manner because there is no linkage to a farm production component.[11] The solution to this simpler problem, where the household maximizes its utility subject to the linearized budget constraint, then yields a standard Marshallian demand equation for leisure that will be of the form

(4.5) $\mathcal{L} = \not b\,(w^*, E^*, S, \Phi)$

where w^* is as defined above and E^* represents the non-labour income equivalent to OD. S is a vector of household and individual characteristics that may affect preferences for consumption and leisure independently of w^* and E^*; and Φ is the labour supply parameter set to be estimated.

The structural labour supply function has the same form as Eq. 4.5 since

(4.6) $L_s = T - \mathcal{L} = T - \not b\,(w^*, E^*, S, \Phi) = \ell\,(w^*, E^*, S, \Phi, T)$

where L_s is total labour supply of the household,[12] and T is conventionally treated as an exogenously given constant.

The two variables (w^* and E^*) needed to identify the linearized budget constraint, and hence to estimate Eq. 4.6, can be derived directly from a prior estimate of a farm production function. As noted before, w^* is just the marginal product of family labour in farm production at the optimum production point. E^* is the level of an assumed endowment income which includes the level of farm profit at the production optimum. This level of farm profit is calculated by deducting all labour costs and other variable costs from the gross value of farm production. The labour costs for family labour are computed on the assumption that family labour is paid a fixed wage rate of w^* for each unit of family labour applied to the farm. That is,

(4.7) $E^* = \pi^*(w^*) + E$

where π^* = shadow farm profit (the maximized value of profit at the equilibrium point, G, using a shadow wage rate, w^*)

$$= \max_{F, H} \{p\, Q(F, H, A) - w^h\, H - w^*\, F\}$$

(4.8a) $= \{ p\, Q(F^*, H^*, A) - w^h\, H^*\} - w^*\, F^*$

(4.8b) $= Y^*_N - w^*\, F^*$.

In Equations 4.8a and 4.8b, F^* and H^* are the optimum choice levels of family and hired labour input in farm cultivation. Y^*_N represents the maximized total returns to farm cultivation net of all purchased inputs, but including the implicit return to family labour applied to own farm work.

Since E^* depends only on variables that are either exogenous or are derived from w^*, the only variable needed to linearize the farm household's budget constraint is w^*.

In summary, the shadow wage method for estimating labour supply for farm households engaged only in self-employment consists of three steps:

i.   estimate a whole-farm agricultural production function in which family labour is recognized as a distinct input;

ii. using the estimated parameters of the production function, derive the marginal product for family labour for each sample household. Using this estimated value of the marginal product as the effective wage rate (w^*) for family labour, derive the shadow farm profit at the optimum production point. Then compute the total non-labour endowment income level E^* as the sum of shadow farm profit and the normal non-labour endowment income E;

iii. estimate a standard labour supply function of the form of Equation 4.6 by regressing total household labour supply on w^*, E^* and S, assuming the household treated these shadow wages and shadow endowment income as being exogenously given to the household *for all levels* of its labour supply. [13]

This methodology of linearizing the concave budget constraint of the farm household works just as well for households participating in the off farm labour market. When $M > 0$ the consumption equilibrium position occurs along a linear section of the household budget constraint (such as PK in Figure 4.1). In this case the household allocates its family labour to own farm production up to the point where the marginal return from farm production is equal to the off farm wage rate (w). Since the wage rate is exogenously given, the equilibrium occurs in an already linear section of the budget constraint. It is still necessary, however, to linearize the entire budget constraint along the equilibrium point. This is required in order to reformulate the farm household utility maximization problem as the simpler utility maximization problem of a consumer household with a fixed non-labour income level and a given labour time endowment.

The linearization is done in the same way as in the case illustrated in Figure 4.1. It is necessary to identify the slope and intercept of a linear budget constraint, which would give the same consumer equilibrium as the concave budget constraint OBGPK when the equilibrium occurs in a linear section such as PK. The slope of the linearized budget constraint will be the market wage rate w (the same slope as PK). The intercept, which gives the imputed non-labour endowment income (E^*), will be computed just as in Equation 4.7 with w^* replaced by w. When $M > 0$, the valuation of family labour in deriving farm profit reflects the opportunity cost of family labour in the off farm labour market. In this case the marginal product of family labour need not be estimated separately to carry out the linearization of the budget constraint. All that is required is to derive the intercept term that represents the value of consumption that can be met from non-labour endowment income, including the imputed value of farm profit.[14]

4.3 A Two Step Estimation Strategy with Heterogeneous Labour

The two step estimation strategy discussed in the previous section is general enough to carry over to the specific framework of a farm household model with heterogeneous labour inputs. In the estimation strategy adopted in this book two adjustments are made to the specific procedures used by

Jacoby (1993). Firstly, the farm production function estimation also embeds a test for the heterogeneity of family and hired labour, using alternative representations of the nature of the efficiency differences between family and hired labour. Secondly, the specification of the labour supply equation is not based on the estimated marginal product of family labour but on the equivalent effective wage rates for family labour that can be derived in a theoretically consistent manner, using the parameters that describe the extent of labour heterogeneity. As described in Section 3.5 of Chapter 3, the effective wage rates for family labour are based on the observed market wage rates for family and hired labour and on the ratio of the marginal products of the two types of labour ($\theta*$), depending on the net labour market position of each household.

Family labour is not modelled as a separate input distinct from hired labour. Instead, a nested production structure is used where aggregate labour is a composite function of family and hired labour. The separable labour composite, if supported by the data, still allows for a general form of labour heterogeneity but it considerably simplifies the analytical solution and estimation of the farm household model. In this case the shadow wage rate for family labour in the labour supply equation need not be set to the estimated marginal product of family labour. One can make use of the additional first order conditions that relate the marginal product of family labour to the market wage rate for *hired* labour on large farms that utilize both family and hired labour. This makes it feasible to derive the correct effective wage rates (w or w^h adjusted by $\theta*$) for family labour, depending on the labour market exposure of the particular farm household.

The first step estimates the production structure implicit in Equation 3.3a (in Section 3.4 in Chapter 3) where family and hired labour form a separable but possibly heterogeneous composite:[15]

(4.9) $Q = f (g (F, H, \Theta), A, V, \alpha) + u$

where A is cultivated land area
  $g(.)$ determines a composite of effective labour in efficiency units, given inputs of family (F) and hired labour (H)
  V represents other production inputs
  Θ is the parameter set of the $g(.)$ function which determines the nature and extent of labour heterogeneity
  α is the parameter set of the production function f(.) with effective labour and other inputs
  u is a random error term.

The test for labour heterogeneity consists of specifying alternative functional specifications for the $g(.)$ function and testing whether the parametric restrictions that lead to a homogeneous labour specification are supported by the data. Labour heterogeneity has two related dimensions:

(i)   Are family and hired labour perfect substitutes in the production function in the sense that the Allen partial elasticity of substitution (AES) between family and hired labour is infinitely large?

(ii)   Are the marginal products per unit of labour time of these two types of labour equal to each other, everything else held constant?

Dimension (i) above is a test for whether the $g(.)$ function is linear. Dimension (ii) is a test for whether the marginal rate of technical substitution (MRTS) between family and hired labour is always equal to one. Several functional forms for the $g(.)$ function are specified in order to test for both dimensions of heterogeneity – i.e. testing for differences in marginal productivity as well as for a well defined AES (whether constant or varying) between family and hired labour.

As discussed in Chapter 3, when F and H can be aggregated in a separable composite given by the $g(.)$ function, the relative efficiency of hired and family labour is determined solely by the parameter set Θ of the $g(.)$ function and the levels of F and H. At the optimal labour allocation let the ratio of the marginal productivity of hired labour *vis-à-vis* family labour be represented by θ^*. With separability,

$$(4.10) \quad \frac{\partial Q / \partial H}{\partial Q / \partial F} = \frac{\partial g / \partial H}{\partial g / \partial F} = h(F^*, H^*, \Theta) = \theta^*.$$

Consequently, the value of θ^* estimated from the production function can be used to derive the effective wage rate and the shadow profit required for estimating the labour supply component of the model in the second step. Instead of defining the shadow wage rate for family labour, w^*, in terms of the estimates of the marginal product of family labour from the production function, Equation 4.6 can be re-specified in terms of the observed market wage rates, w and w^h, and θ^*, (and by using Equation 4.7), as follows:[16]

$$(4.11) \quad L_S \quad = \quad \ell\left(w^*(\theta^*), \pi^*(w^*) + E, S, \Phi, T\right) + e$$

where $w^*(\theta^*) = w$ if $M > 0$ (small farm)

                 $= w^h/\theta^*$ if $H > 0$ and $M = 0$ (big farm)

$$= \omega \qquad\qquad \text{if } H = M = 0 \qquad \text{(autarchic farm)}$$
$$\text{where } w \leq \omega \leq w^h/\theta^* \quad \text{for } \theta \leq 1.$$

$$\pi*(w^*) = \; Y^*_N - wF \qquad\quad \text{if } M > 0 \qquad\qquad \text{(small farm)}$$
$$= \; Y^*_N - (w^h/\theta^*) \, F \quad \text{if } H > 0 \text{ and } M = 0 \;\; \text{(big farm)}$$
$$= \; Y^*_N - \omega F \qquad\quad \text{if } H = M = 0 \qquad \text{(autarchic farm)}$$

and $\quad Y^*_N \quad =$ the optimized value of net farm income including returns to family labour. (See Eq. 4.8a and 4.8b).

$\quad\quad\;\; S \quad\;\; =$ a vector of household and individual characteristics that affect preferences between leisure and consumption

$\quad\quad\;\; e \quad\;\; =$ random error term.

The advantage in the estimation structure of Equations 4.9 and 4.11 is that it avoids direct use of the estimated marginal product of family labour in the labour supply equations.[17] Predicted values of the marginal product of family labour at the household level are likely to show an extreme level of variation which may not be consistent with the variation in the unobservable underlying subjective valuation of family labour. Unless certain restrictive functional forms are used, the estimated marginal product of family labour could also be negative in a large percentage of the sample households from which the production function is estimated. In Jacoby's own exercise, when the production function was estimated as a fully specified translog equation, nearly 20% of the sample households produce a negative marginal product for female family labour.[18]

Another problem with using marginal products of labour in the labour supply equation to derive both the shadow wage and the shadow farm profit is that it increases the likelihood that errors of the production function and labour supply equations (u and e) will be correlated. Such correlations increase the complexity of the two step estimation procedure and of the required adjustments in the standard errors of the parameters estimated in the second step. While the potential endogeneity of the marginal product based measures of w^* and E^* in Equation 4.11 could be addressed by using instruments for the shadow wage rate and shadow farm profit, the choice of instruments that are not only uncorrelated with e but also uncorrelated with u is likely to be limited.

The two step sequential estimation strategy proposed above – where some of the regressors in the labour supply equation are derived from parameters estimated in the production function equation – is an example of what Pagan (1984) refers to as a two step procedure for estimating models with a "generated regressor". In these models the main interest is in obtaining consistent estimates of the parameters in the second stage regression when it contains variables that, while directly unobservable, are estimable from the parameters of an auxiliary equation estimated in the first step.[19] The interests in this study, however, differ slightly from the usual model with generated regressors. The auxiliary equation of the production function serves not only to generate the unobservable variables for the labour supply equation but also to detect the nature of the heterogeneity between family and hired labour, as reflected in the Θ parameter set.

An alternative strategy to the two step procedure outlined above would be to estimate the first (auxiliary) and second step models via some joint method, such as full information maximum likelihood (FIML), as in Liederman (1980). In reference to Equations 4.9 and 4.11, FIML estimation would imply that the parameter vector Θ, reflecting the heterogeneity between family and hired labour, would be estimated jointly from the production function and the labour supply equations. Our tests for labour heterogeneity would likely be more robust if the estimates of the Θ parameter set were derived from a joint estimation of the production and the labour supply function. In addition there would be the usual efficiency gains if the error terms in the two equations were to be correlated. A joint estimation strategy for Equations 4.9 and Eq. 4.11, however, is equivalent to joint estimation of the production and consumption components of the underlying farm household model. As noted in Section 4.1, such a joint estimation strategy for farm household models is computationally complex; and, depending on the nature of the heterogeneity implied by the Θ parameter set, it may not always be feasible to solve analytically for the reduced form equations. If so, one cannot take advantage of economic theory in imposing or testing for parametric restrictions in the estimated equations (Singh, Squire and Strauss, 1986c1: 21).

Another reason why FIML is not used in this study is the structure of the data. The farm production function is estimated at the level of a household while the labour supply regressions are done with data on individual household members. With such a data structure it is awkward to hypothesize the joint distribution for the error structure in the first and second step regressions. While the labour supply regressions could also be estimated at the household level by averaging over individual members, it is a better option to work with the individual level data since individual characteristics can be important in determining labour supply behaviour in addition to shared household level characteristics.

While opting for the sequential estimation procedure in which the Θ parameter set is derived from the production function equation only, additional sensitivity analysis is carried out in Chapter 6 to show that the production function based results on labour heterogeneity are robust to many alternative assumptions and specifications of the production function. Furthermore, while the labour supply equations do not directly contribute to the estimates of Θ, they can be used to provide independent corroboration of the nature of labour heterogeneity detected in the production function estimation. This is accomplished through appropriate model selection diagnostic tests reported in Chapter 7. It is possible to compare model specifications that allow for a common effective wage (which is consistent with family and hired labour being homogeneous inputs) with model specifications that use varying effective wages, conditional on the estimated Θ (which is consistent with labour heterogeneity).

4.4 Error Correction for the Two Step Estimator

The basic strategy of the two step procedure is to replace the unobserved regressors in the second step with their estimated or predicted values from the auxiliary or first step regression. These values are then treated as if they were known *a priori* for the purpose of estimation and inference in the second step model. That the two step procedure yields consistent estimates of the second step parameters under fairly general conditions is well documented, as is the fact that the second step standard errors and related test statistics reported in the normal regression output are normally incorrect (Pagan, 1986; Murphy and Topel, 1985). However, the need to correct the standard errors in the second step is commonly ignored.

The standard errors in the second step require an upward adjustment to account for the fact that the generated regressors are measured with some error that is related to the precision of the parameter estimates in the first step. In an illustrative example reported in Murphy and Topel (1985: 372), the proportional adjustments required for the estimated standard errors are largest for the generated regressors. The required adjustments are somewhat smaller for variables that appear in both the first and second step equations, and negligible for the non-generated exogenous regressors in the second stage which do not also appear in the first step equation.

In the specific framework of the model estimated in this study the key parameters of interest in the second step equation are the wage and income elasticities of labour supply. Since the wage and (non-labour endowment) income variables are the generated regressors in Equation 4.11, for which the adjustments in the standard errors of the relevant parameters are likely to be the largest, it will be important to make the required adjustments.

The procedure adopted for adjusting the second step standard errors follows that suggested by Murphy and Topel (1985). These adjustments are relatively straightforward when the second stage equation is linear as in Equation 4.11, and when it is further assumed that the random errors of the first and second step equations are uncorrelated.[20] This latter simplifying assumption is also adopted in Jacoby (1993).

The general structure of a two step estimation procedure is written as:

(4.12) $q = x_1 \alpha + u$

(4.13) $y = x_2 \beta + f(\alpha, x_1) \gamma + e$

where α is a vector of unknown parameters estimated in the first step, based on a vector x_1 of exogenous variables which determine q; and y is the variable of interest in the second step regression, with y being influenced both by an exogenous set of variables, x_2, and another set of variables which are actually unobservable but which can be generated with the help of the α parameters (and the x_1 variables) of the first step equation. These unobservable variables are given by $f(\alpha, x_1)$. The set x_1 and x_2 could, of course, overlap and f(.) need not be related in any way to the function that determines q from x_1. The only restriction imposed on f(.) is that it be twice continuously differentiable for each α in x_1. Hence f(.) could be non-linear in α.

It is assumed that the first step regression yields an estimator $\hat{\alpha}$ of α which is consistent and asymptotically normal with variance-covariance matrix $V(\hat{\alpha})$. Let $\hat{V}(\hat{\alpha})$ be a consistent estimate of $V(\hat{\alpha})$ obtained from the first step regression. In the second step y is regressed on x_2 and the estimated values $f(\hat{\alpha}, x_1)$; and what is required is the correct asymptotic distribution of (β, γ) for tests of statistical significance and other inference based on a least squares regression of Equation 4.13.

Let r be the number of elements in the $\hat{\alpha}$ vector and n the number of observations for the second stage regression. Denote by X_2 be the $n \times p$ matrix of observations on x_2, and let B be the $n \times m$ matrix of generated regressors $f(\hat{\alpha}, x_1)$. Denote the full observation matrix of the second step equation as $Z = (X_2, B)$. The covariance matrix for (β, γ) when Equation 4.13 is estimated by ordinary least squares is given by:

(4.14) $\Sigma_0 = s^2 (Z'Z)^{-1}$

where s^2 is the sample estimate of the variance of the error term e.

The correct asymptotic covariance matrix for (β, γ), adjusted for the fact that $\hat{\alpha}$ is estimated with error, is given by [21]

$$(4.15) \quad \Sigma = \Sigma_0 + (Z'Z)^{-1} Z'B^* \hat{V}(\hat{\alpha}) B^{*'}Z(Z'Z)^{-1}$$

where B^* is computed as the derivative matrix of B with respect to α, such that the typical element b^*_{ij} of B^* is given by [22]

$$(4.16) \quad b^*_{ij} = \sum_{k=1}^{m} \gamma_k \frac{\partial f_k(\hat{\alpha}, x_{1i})}{\partial \alpha_j}.$$

The form of Equation 4.14 indicates the error correction procedure for the second step equation exceeds the commonly reported covariance matrix Σ_0 by a positive definite matrix (given by the second term of Equation 4.15). As a result, the standard errors based on the unadjusted Σ_0 in a naive two step procedure are always understated. It is also clear the size of the adjustment in the standard errors of (β, γ) depend crucially on two factors: $\hat{V}(\hat{\alpha})$ which is the precision of the first step parameter estimates, and the correlation between the second step explanatory variables Z and the derivative matrix B^*. To the extent that this correlation is high, or the sample variance in the first step regression is large, the error adjustment for the second stage will be more important (Murphy and Topel, 1985: 375).

4.5 Summary

The estimation methodology for this study follows a two step procedure adapted from the method proposed by Jacoby (1993) for estimating non-recursive farm household models. In the first step, an aggregate farm production function, containing a separable labour nest, is estimated and used to test for the heterogeneity between family and hired labour. Several alternative specifications that allow for imperfect substitution between family and hired labour and unequal marginal products are estimated. In the second step, a structural labour supply equation is estimated with a specification that is consistent with the type of labour heterogeneity found in the production function estimation. This typically means that the effective wage and non-labour income variables used in the labour supply regression will be derived from parameters estimated in the production function.

Jacoby's approach was to base these generated regressors of the labour supply equation on the marginal product of family labour estimated from the production function. In Chapter 3 it was shown that the marginal product for family labour at the optimum labour allocation could be directly related to the observed market wage rates for family and hired labour, when the heterogeneity between family and hired labour is explicitly modelled within a separable labour nest in the production function. Depending on the net labour market position of specific households, it is possible to derive the effective wage rate that determines their labour supply equilibrium from the observed market wage rates and the parameters that reflect the labour heterogeneity. This provides an alternative specification of the labour supply equation that has several advantages over the specification which directly uses the estimates of the marginal product of family labour at the household level.

An alternative estimation strategy based on full information methods to estimate the production and labour supply equations jointly would have allowed for the nature of the labour heterogeneity to be simultaneously determined from the production and labour supply behaviour of the farm households. The complexities of such joint estimation, however, are formidable. The method adopted for this study is the simpler but widely used alternative of sequential estimation which gives consistent estimates under quite general conditions, requiring only a straightforward adjustment to the standard errors of the parameters estimated in the second step.

Notes

1   The full income of the farm household includes the imputed short run farm profit that represents the pure returns to the farm household from its ownership of the land input. (See Equation 3.2n in Section 3.2 of Chapter 3). The method of imputing the value of farm profit already reflects the labour supply choices of the farm household. Hence, the consumer demand side estimation is conditional on this specific level of full income.

2   The recursive structure of the farm household model may break down for other reasons as well - such as, if home produced consumer goods are imperfect substitutes for market purchased consumer goods, or if the household assigns different levels of dis-utility to workdays applied to own farm cultivation and workdays on the off farm labour market. Lopez (1984) is an example of a non-recursive model based on varying preferences between work on the family farm and off farm work.

3   The subsequent discussion in this paragraph closely follows Singh, Squire and Strauss (1986c2: 48).

4   For instance, consider the comparative statics of a change in the price of farm output on household leisure demand. In the recursive model with a fixed wage for family labour, the output price change has only an income effect because it changes the level of farm profit. In a non-recursive model with a subjectively determined shadow wage rate for

family labour, the change in the output price has an additional effect on leisure demand. The household-specific shadow price for family labour that equates family labour supply and demand will also adjust in response to the price change. The latter term - the marginal effect on the shadow wage for family labour due to a change in the output price - will not be estimated in a recursive model structure. See Strauss (1986) for the comparative statics of a farm household model involving shadow or virtual prices.

5   Burtless and Hausman (1978) and Blomquist (1983) are early examples.

6   Since the price of the consumption good p_c was normalized to one, the value of farm output can be incorporated directly into the household's consumption budget constraint without distinguishing between real and nominal consumption quantities.

7   This avoids corner solutions involving complete non-participation of family labour in productive activity. If the separate labour force participation decision is of interest, it can be modelled in alternative ways - for instance, with a probit model.

8   Figure 4.1 is identical to Panel A in Figure 3.1 of Chapter 3, except for the extension of the vertical axis to include the OB segment of the budget constraint. Point G in Figure 4.1 is identical with point G in Figure 3.1 Panel A.

9   In Figure 4.1 the full range of the non-linear budget constraint OBGPK for a farm household which does not supply any labour on the off farm market can be represented by a single linearized budget constraint (DN) at the optimum point. This is in contrast to the linearization process in the presence of progressive income taxes which results in several piece-wise linear budget constraints. The analytical method to solve for the equilibrium labour supply position and the econometric procedures for estimating labour supply functions with piece-wise linear budget constraints is considerably more complex (Moffitt, 1986) than in the case of the farm household model outlined here.

10 λ is the Lagrangian multiplier associated with the farm household utility maximization problem given in Chapter 3 by Equation 3.6, and μ_1 is the complementary slackness parameter associated with the non-negativity of off farm labour supply (the M variable) as specified in Equation 3.7a.

11 The only difference between the textbook version of a labour household equilibrium and the farm household equilibrium with a linearized budget constraint is in the interpretation of the full income of the household. In the former the full income is completely determined by E and w*T and does not reflect any household choices. In the latter the full income of the farm household is conditioned on the observed demand for leisure, and conversely the optimal leisure/labour choices are also conditioned on the production choices determining the level of farm profit (Jacoby, 1993: 906).

12 L_s includes all components of the household labour supply, including other productive activities, such as home processing and cottage industry related employment. At the optimum, the marginal returns to all productive activities should be set equal to each other. It is not required that the shadow wage rate be defined only in terms of the marginal product of family labour in farm production. The marginal returns to other household production processes could be used as well, although it would normally be more difficult to estimate the production function for these other activities.

13 In the actual estimation it would be necessary to use an instrumented version of w* and E* since in a statistical sense these variables cannot be assumed to be uncorrelated with the errors in the labour supply equation. Part (iii) can be estimated independently or as part of a joint demand system with the other consumption goods of the model. This depends on whether leisure is assumed to be separable from other goods in the utility function (Deaton and Muelbauer, 1980). Alderman and Sahn (1993) is an example of a joint estimation of leisure and commodity demands using an AIDS specification.

14 The linearized budget constraint when family labour is applied both to own farm cultivation and to the off farm labour market can also be readily illustrated with reference to Figure 4.1. In this case the production equilibrium will be given by the tangency of a wage line with slope w with the production function surface BGPQ. To derive this point of tangency within Figure 4.1, simply shift upwards the wage line DK, with slope w, till it is tangent to the BGPQ surface. Let this line of tangency with slope w be represented by D'K' and let OD' be its intercept on the consumption axis. Let P' be the point where D'K' is tangent to BGQ. (D'K' and P' are not illustrated in Figure 4.1). P' must be to the right of the point G as drawn in Figure 4.1 because the tangency line DN at point G has a higher slope by assumption. D'K' represents the linearized budget constraint for the farm household which works on the off farm labour market. Its consumption equilibrium will be represented by the tangency of the D'K' line with the highest level indifference curve. This point will be further to the left than P' because by assumption the total labour supply of the household exceeds the family labour applied to own farm cultivation when M > 0.

15 There is a prior step involved in which it is necessary to test whether F and H can indeed be represented in a separable composite given by the $g(.)$ function. These tests for input separability are described and carried out in Chapter 6.

16 Assuming $w < w^h/\theta^*$. Otherwise all desired family labour is supplied to the off farm wage market by all household categories, and the effective wage rate is w for all individuals. See the discussion in Section 3.5 of Chapter 3.

17 In the specification of Equation 4.11, an estimate of the marginal product of family labour is required only to exactly identify the effective wage rate of the autarchic household since in equilibrium ω will equal the marginal product of family labour. The alternative specification based on the observed market wage rates w and w^h and the θ^* value, however, can still be used to define a fairly narrow range over which the effective shadow wage rate must lie even in the case of the autarchic household.

18 Jacoby addresses this problem by dropping specific interaction terms involving female family labour that have a negative coefficient in the fully specified translog equation with some loss of generality in the flexible functional form. (Jacoby, 1993: 913 footnote 14).

19 Models in which generated regressors occur are widely used in econometric applications. Some early example are models of (rational) expectations where the variable reflecting the anticipated value or expected value is generated as the predictor from another equation representing the expectations process (Barro, 1977; Topel, 1982); and spatial models of labour market equilibrium which include as a regressor the predicted probabilities for unemployment (Abowd and Ashenfelter, 1981).

20 Under these conditions the standard error adjustments can be made on the basis of the normal regression output of most econometric packages. When the errors are correlated the adjustments are more complicated. See Newey (1984) for details on the required adjustments with correlated errors.

21 Murphy and Topel (1985: 374-375).

22 In Equation 4.16 the index i represents the observation number (from 1 to n), the index j ranges over the number of elements in the α parameter set (from 1 to r); and k indexes the number of generated regressors in Equation 4.12 denoted by $f(\hat{\alpha}, x_1)$, which by assumption is 1 to m. Hence the matrix B* is $n \times r$ while B is $n \times m$.

5 The Setting and the Data

This chapter has three parts. Section 5.1 contains a cursory description of the southern lowland region of Nepal (referred to as the *tarai)* that is the setting for the survey data used in the farm household model estimation. It also briefly explains why the question of heterogeneity between family and hired labour in Nepalese agriculture is particularly relevant in the context of the *tarai* region. Section 5.2 describes the design and main features of the household survey and the structure of the data collected. The last section discusses the definitions adopted for the main variables used in the production function and labour supply regressions, and shows how they were computed from the data recorded in the household survey forms.

5.1 The Setting

Nepal has three distinct geographical or ecological regions that lie on a north to south axis throughout the whole country. They are the northern most mountainous region, the middle hill region and the southern lowland *(tarai)* region. The material conditions for agricultural production are very different in the three regions. These differences derive mainly from climatic factors related to variations in altitude. Over the years such differences in physical conditions have led to very distinct regional agrarian structures and the social relations of production.

The empirical analysis of this book is based on household level survey data from the *tarai* region of Nepal. The data are drawn from a large nationally representative household budget survey carried out by Nepal Rastra Bank (the central bank of Nepal), called the Multi-purpose Household Budget Survey (MPHBS). The MPHBS collected household data from a representative sample for each of the three ecological regions of Nepal, but the complete data set is not utilized in this study. The farm household model estimation uses the data from the *tarai* sample only because the question of heterogeneity between family and hired labour in Nepalese agriculture is best addressed in the context of the *tarai* region.

The MPHBS sample of households from the northern hill and mountain districts of Nepal has been excluded from this study mainly because of the very limited use of hired labour in those regions. The average farm size in these northern regions is only about 0.7 hectares per operated holding (HMG/N/CBS, 1993). Farm production is subsistence oriented and only a limited form of economic differentiation occurs among farm households in these regions, most of who are owner-cultivators.

The agrarian structure in the southern *tarai* region of Nepal more closely follows the classical structure of a few large landlords dependent primarily on hired labour, a middle group of owner-cultivators and a large group of landless households who supply the hired labour on the larger farms. Although the average farm size in the *tarai* region is still small (about 1.2 hectares), a wider socio-economic differentiation of the rural households occurs. A significant percentage of the land is cultivated in operated holdings in excess of 10 hectares; while on the other extreme, a significant proportion of the population consists of landless households who work as hired labourers on the big farms.

The National Sample Census of Agriculture for Nepal gives data on the size distribution of land holdings based on operated area and not on owned area. So there are few completely landless households enumerated in these censuses since those who do not own any land usually operate small plots on lease or as sharecroppers. On the basis of the operated size of farm holdings from the 1991/92 Census,[1] more than 10% of the *tarai* region households operate less than 0.1 hectares of land. Another 8% operate between 0.1 and 0.2 hectares. On the other hand, while holdings greater than 10 hectares comprise only 0.5% of the total holdings (households), they account for about 7% of the total operated land.

The rural labour markets in the *tarai* region are quite active. In peak periods there is considerable in-migration of agricultural labourers from India as well (Wallace, 1989). Because of better regional transportation infrastructure, local villagers also have easier access to off farm wage employment opportunities in other villages and nearby urban centres. Rural households in the *tarai* region of Nepal are less likely to be quantitatively constrained on either their labour demand or labour supply choices than farm households in the northern hill/mountain regions.

A distinctive feature of hill region farming is the arrangements for reciprocal labour exchange among household groups. The incentive structures for exchange labour will differ from that for hired as well as for own family labour. This is another reason why the present study focuses on the *tarai* region only, where exchange labour arrangements are quite rare.

Nepal's *tarai* region is an extension of the Indo-Gangetic plains of northern India. Within Nepal the *tarai* comprises a narrow strip of land along the southern border at relatively low altitudes in comparison to the hilly north. *Tarai* region crop production is undertaken mainly on level land, termed *khet*, consisting of alluvial soil, which can be irrigated with the monsoonal flow of the region's rivers. Farms in the northern areas consist mainly of terraced hillsides that are difficult to irrigate.

The *tarai* region has about half of the population of Nepal and two-thirds of the cultivated land. The main summer crop is paddy, sometime grown in two rotations where irrigation is available. Wheat, pulses and oilseeds are the main winter crops. Maize and millet are the common rotation in upland plots, termed *pakho*, not suitable for paddy cultivation.

Agricultural production technology in the *tarai* region of Nepal is still quite backward. Average paddy yields are only about 2 metric tons per hectare and have been declining (HMG/N/CBS, 1997) as more and more marginal land is brought under cultivation. Human labour is the main farm input, typically accounting for more than 50% of the total cost of farm cultivation.[2] There is a significant east-west sub-regional variation within the *tarai* belt of Nepal. The monsoon rainfall decreases as one travels from east to west and so does population density as a consequence.

The sub-regional economies of the *tarai* region are more closely integrated with the Indian border markets than with each other (Wallace, 1989). Hence one observes considerable variation in output and input prices within the *tarai* region of Nepal. This variation extends to market wage rates for agricultural labour. The potential variation in wage rates and other prices is an important feature of the MPHBS data for the *tarai* region sample. This variation makes it feasible to estimate the labour supply regressions with a single cross-sectional data set for the *tarai* region only.

5.2 The Data Set

5.2.1 Source

The MPHBS was conducted by Nepal Rastra Bank in a phase-wise manner in 1984 and 1985, and its detailed farm household related data was collected with reference to the 1984/1985 annual cropping cycle. It is the most extensive of the periodic but irregular household budget surveys conducted by Nepal Rastra Bank. These surveys are designed principally to compile information for the preparation of consumer price indices. The

MPHBS, however, had a more ambitious objective of improving the Nepalese database on the regional patterns of employment, income distribution and consumption because regular surveys of this nature are not conducted by other government agencies in Nepal.[3]

The full sample size for this survey was 4,022 households. The rural component consisted of 3,660 households, drawn from a stratified sample of 22 districts (out of a total of 75 in Nepal). The MPHBS data used in this study consists of a subset of the rural sample from the southern *tarai* region of Nepal. The total sample of the *tarai* region in the MPHBS was 1,571 households from seven sample districts. The estimation work reported in Chapters 6 and 7 is based on a smaller subset of the *tarai* region sample, consisting of 1,007 households, which is the complete household sample from five out of the seven *tarai* districts of the MPHBS.

The reason for the further reduction of the *tarai* sample size is that not all components of the MPHBS data used in this study was made available from Rastra Bank in computerized form. Because of the vast amount of data collected and the extensive work required for processing and data cleaning, some parts of the MPHBS questionnaire were not computer processed at Rastra Bank. Unfortunately, one component of the survey left out of the Rastra Bank data tapes was the extensive data collected on the farm management module of the survey with the details of the crop-specific farm outputs and production inputs.

The processing of the farm management data component of the MPHBS was separately organized by the author at the Institute for Integrated Development Studies (IIDS), Kathmandu, from the original survey questionnaires. At IIDS it was not feasible to process the complete farm management component from all of the 3,660 rural household questionnaires. An arbitrary choice was made to process the full household sample from only 13 of the 23 sample districts, based on regional representativeness. The additional data processing at IIDS was initially done for other purposes than the research work reported in this book. The *tarai* region coverage of the IIDS data component is the full sample of households from five of the seven selected *tarai* districts of the MPHBS. This represents close to two-thirds of the original MPHBS sample size for the *tarai* region (1,007 out of a possible total of 1,571 *tarai* region sample households). The five *tarai* districts included in the IIDS data subset are Morang, Mahottari, Rupandehi, Banke and Kailali.[4] (The location of these sample districts is indicated in Figure 1.1 in Chapter 1). Each of these districts represents one the five regional development zones of Nepal, moving from east to west, respectively.

The production function estimates reported in Chapter 6 are based primarily on the farm management data subset of the MPHBS that was processed at IIDS. This component of the survey has been linked with the computerized data files obtained from Rastra Bank on other parts of the survey to prepare a fully integrated database of the 1,007 *tarai* region sample households for the labour supply regressions of Chapter 7.

In spite of the long period since the data was collected, the MPHBS remains one of the most detailed household level survey data collected for Nepal.[5] It was well designed and implemented, and the data collection was carefully monitored. Information was collected over several rounds. One of its special features was to collect data on many non-monetized transactions (i.e. income received in kind as gifts and transfers, or as a part of wages) that are important sources of income in poor developing countries but are often overlooked in household surveys.[6] In spite of its richness, the data from the MPHBS has been rarely used for subsequent research work on Nepal. This neglect provided part of the motivation to base the estimation work of this study on the MPBHS data.

5.2.2 Sample Design

The sample design of the MPHBS was based on a four step selection procedure. In the first step, 22 sample districts were chosen, based on random sampling from a regional stratification of all the districts of Nepal.[7] Within each sample district the subsequent steps followed in selecting the final sample of rural households consisted of:

ii.   a random selection of sample village *panchayats*;
iii.   a random selection of 2 (out of 3) clusters from each selected *panchayat*, reflecting a grouping of the administrative sub-divisions within a *panchayat*;[8]
iv.   a stratified sample of households from each cluster, selected from five household strata based on the size of the household's operated land.

The *panchayats* selected in step (ii) correspond to the main local administrative units within a district.[9] The *panchayat* boundaries in rural Nepal are drawn to include about 1,000 to 1,200 households, and so they do not usually coincide with a single village. Nonetheless, the geographical area covered within a single *panchayat* tends to be homogenous, particularly in the *tarai* region which has a much higher population density than the northern hill regions of Nepal. Hence, for convenience of language, the term "village" is used interchangeably with "village *panchayat* " in the rest of this chapter and elsewhere in the book.

In the last step of the sample design there was a complete enumeration of all households within the selected clusters. These households were then classified into five sampling strata (household classification categories) based on the size of the operated land of the household. Operated land consists of owner-cultivated land and land rented in. The first four strata defined households with operated farm sizes designated as marginal, small, medium or large. A fifth stratum consisted of landless households. Within each stratum a five percent random sample of households was drawn, with an upward adjustment in the sampling ratio where necessary to ensure that at least two households were selected from each stratum.[10]

The operated land holding strata limits adopted in the sample design of the MPHBS for the *tarai* region are as follows: marginal farm holdings are up to 1.02 hectares, small holdings are up to 2.73 hectares, medium holdings are up to 5.44 hectares, and large farm holdings are above 5.44 hectares. Households with only a homestead that may have included a kitchen garden were classified as landless.

The important implication of the MPHBS sample design is that within each sample village *panchayat* the material conditions of farm production, the cropping pattern and wage rates and prices will tend to be fairly uniform. But these will vary considerably across *panchayat* and districts, especially since the five district sub-sample has been selected from the entire range of the east-west regional variation in Nepal. Within a village *panchayat* (and consequently within the selected sample clusters), the main basis for the variation in the household level data will be with respect to income and assets holdings based on the amount of operated land. Hence, the MPHBS data is a broadly representative sample from different socio-economic strata within a village. Such a variation proves useful in the labour supply estimation work to identify the income effects across households that face more or less equal wage rates.

The final sample of 1,007 households used in the empirical estimation of this study does not necessarily constitute a completely random sample for the *tarai* region of Nepal. Within each of the five selected *tarai* region districts, the sample of households used in this study is the complete set of households chosen from these districts in the MPHBS sample design. So this can be treated as a random sample of households within each of these districts (subject to the stratified clustered sample design described above). However, the random sampling properties of the overall *tarai* sample of the MPHBS may not carry over to this data subset drawn from only five of the seven *tarai* districts selected in the full MPBHS sample. In the regression estimations of this book no specific attempt is made to deal with this problem by modifying the weights given to the households to reflect the *ad hoc* selection of the five *tarai* districts. The possibly unrepresentative nature of the selected *tarai* region data subset, however, is not a major issue because the focus of this study in not on estimating *tarai* region aggregates or household means.[11]

The distribution of the final sample of 1,007 households used in the estimation work of this book is given in Appendix Table 5A.1, classified according to sample district and land size strata.

5.2.3 Farm Management Data Structure

The farm management module of the MPBHS collected detailed data on all farm outputs and inputs over a multi-round annual survey period. The farm household's own-account production activities were broadly classified into agricultural and non-agricultural enterprises. The main sub-categories identified under agricultural enterprises were crop production, animal husbandry, horticulture and fishery. Crop production is the dominant farm activity. For the full *tarai* region sample of the MPHBS, the crop production component constitutes more than 81% of total agricultural enterprise income; and agricultural enterprise income in itself accounts for about 65% of total income for the average *tarai* region rural household.[12]

The MPHBS collected very detailed input data at the level of individual crops. In particular, the total labour input applied to a specific crop was classified into nine categories based on three types of workers (adult male, adult female and child) and three sources (family, hired and exchange labour). Detailed information was collected on the actual wage payments and meal expenses related to hired and exchange labour used by a household for a specific crop. So an average wage payment for each type of labourer can be computed for every household reporting use of hired and exchange labour for a specific crop. Data on other production inputs such as bullock days and seeds are also distinguished by source – whether hired (or purchased) or from the household's own resources. In almost all categories the information is collected in both quantity and value terms, with the main exception being chemical fertilizer.[13]

The farm management data set is organized into a winter and summer cropping cycle/rotation. This structure can be used to derive average wage rates paid for hired labour for crop production in each of the two cropping cycles to get a measure of seasonal variation in the wage rate.

Although the farm management survey data, as with the rest of the MPHBS information, was collected through recall based interview methods, several steps were taken to ensure reliability and accuracy of the farm management data. For instance, the data on crop output was cross tabulated with a table on the disposition of farm output (market sales, home consumption, remaining stocks) to produce a household balance sheet for each crop. Similarly, the information on family labour input in crop production was crosschecked with the labour supply component of the survey questionnaire to maintain consistency.

Detailed data were also collected on the size and type of land operated by each sample household. The household level land holding data is also distinguished by land tenure forms: owner cultivated, leased out and leased in either as fixed rent or share cropping arrangements. The main distinction in agricultural land in the *tarai* region of Nepal is between wet land (*khet*) suitable for growing paddy and sloping upland (*pakho*) that does not retain water for paddy cultivation. Within each category a further distinction is made between irrigated and unirrigated land. This makes it relatively easy to control for the differences in the quality of the land input. The cropping pattern between *khet* and *pakho* land tends to be very different. Hence defining the land input to take account of the heterogeneity between *pakho* and *khet* land is one way to account for differences in the crop composition across farm households. While this information gives a good indication of the aggregate quality of the land input available on each farm, the area planted to each crop, unfortunately, is not broken down into these specific land types. The crop level land input variable is recorded as a single value for the total area harvested.

5.2.4 Labour Supply Data Structure

The family labour supply component of the MPBHS data is also reported in great detail. A complete enumeration of all household members with details on individual characteristics (age, marital status, education level, etc.) was prepared in the first survey round. Members of the household who are part of the extended farm family are clearly distinguished from temporary residents, servants and other live-in hired help. The usual occupation for each household member is indicated on the basis of a very detailed three digit coding structure, and this information is used to determine whether a particular person is economically active or not. For each economically active person (aged 10 or above) data on the seasonal employment pattern was collected through multi-round interviews. These data were then aggregated into total days of productive work in each month of the survey year. The main categories of employment distinguished for the number of days worked in each month are: work on own family farm, work on the hired labour market and work as an exchange labourer.

A major shortcoming of the labour supply data set is the missing information on the wage rate received by each individual working in the hired labour market.[14] Nor does the data distinguish the place and type of employment – i.e. whether it is agricultural labour or non-agricultural employment, and whether it is within the sample village or outside. This deficiency means the wage rates for individual family members who report working on the off farm labour market must be imputed from some other part of the MPHBS data set.

5.3 Main Variable Definitions

5.3.1 Production Function

The production function estimations reported in Chapter 6 are based only on the crop production component of the farm management data, which is the dominant part of farm production in the *tarai* region. The other components of agricultural production identified in the MPHBS – animal husbandry, horticulture and fisheries – are ignored. These other activities are ignored because they are minor components of farm production; and because detailed input data are not consistently reported for these other farm activities in the MPHBS.[15] Secondly, there is very little use of hired labour in these ancillary activities, and so they are not relevant to an estimation framework where the main objective is to test for the heterogeneity between hired and family labour.

Although the MPHBS contains data on crop-specific farm inputs and output, the test for labour heterogeneity carried out in Chapter 6 is embedded in an *aggregate* production function for all crops mainly because the important information on land quality is not available at the level of individual crops. Additionally, the differences in the efficiency of family and hired labour may also lead households to choose alternative cropping patterns based on the labour requirements of different crops. One particular manifestation of the labour efficiency related production choices made by households could be more intensive multiple cropping. Such responses will not be fully captured in comparisons of labour productivity in single crop production functions. Furthermore, since the family labour supply data are also at the aggregate level (i.e. not distinguished by crop-specific workdays) it makes sense to derive the effective wage rate for family labour in terms of an average comparison of the productivity of family and hired labour, rather than crop-specific comparisons which may vary considerably across crops.

On the other hand, working with aggregate farm output does create potential bias in the estimation results when the crop composition and output prices are not uniform across different households and regions (Bardhan, 1973). The price variation problem can be accounted for by measuring farm output as a composite quantity variable rather than simply as the market value of total output. The quantity index for aggregate crop output is computed by deflating the total value of farm output by a village-specific aggregate price index. The village level price indices are derived as a Tornquist (log-linear) price index of individual crop prices, using as weights the share in total value of different crops grown in that particular sample village.

From Equation 4.9 in Chapter 4, the specification of the production function with a separable nest for labour inputs is of the form:

(5.1) $Q = f (g (F, H, \Theta), A, V, \alpha) + u$

where Q is aggregate composite output as described above.

On the input side, in addition to land (A) and composite labour (represented by the $g(.)$ function), two other main input variables were created from the MPHBS survey data to represent the set V. They are bullock-pair workdays (B) and material inputs (M).

The material input variable simply lumps together several diverse production inputs identified in the MPBHS data in value terms. M is the sum of the value of seeds, chemical fertilizers, insecticides, irrigation charges, and operating cost and rental charges for farm machinery. The bulk of M, however, consists of the value of seeds and fertilizers since only a small percentage of farms reported machinery use or payment of fees for irrigation water. M is converted into real units by deflating the nominal sum by the village-specific crop price index. Such a deflator is justifiable since the major component of M is the value of seeds.

Apart from the components of M, all of the other major inputs are already reported in quantity levels in the MPBHS data. The main computations required for defining these variables are aggregation procedures over sub-categories.

As noted above regarding the labour inputs, the MPHBS reports workdays on the family farm by male adult, female adult and child labour under each of three sources of labour: family, hired and exchange. The input of child labour is not broken down further by gender. Fortunately, the overall level of child labour reported as inputs in crop production is minimal in the MPHBS farm management data for the selected *tarai* districts.[16] Hence in this study the child labour category is completely ignored in creating the sub-aggregates of total family and hired labour. The level of exchange labour in the *tarai* region is also minimal. Where it occurs it has been treated as hired labour.[17]

The reported workdays of adult male and female labourers are added together after simply converting the female workdays into equivalent male days, based on the observed ratio of the female and male adult wage rates for hired labour in a particular village. That is, total family labour days (F) is measured in terms of adult male equivalent units and is computed as $F = F_m + (w_f / w_m) F_f$ where the m and f subscripts signify male and female sub-categories. A similar procedure is used for aggregating hired labour.

This specification treats male and female labour as perfect substitutes, adjusted for a constant productivity difference represented by (w_f / w_m) – the ratio of the observed female and male hired labour wage rates in a sample village. This is admittedly a crude manner of handling any heterogeneity between male and female labour (within both the hired and family labour sub-aggregates). It is, however, theoretically justifiable to the extent that differences in competitively set market wage rates are likely to reflect differences in labour productivity at the margin.[18]

An alternative treatment of the potential heterogeneity of male and female labour was attempted by distinguishing four different labour inputs in the labour aggregator $g(.)$ function. Using such a structure with the flexible functional specifications of the production function proved intractable. Instead of persisting with an awkward four input $g(.)$ function, this study has opted for the alternative approach of carrying out detailed sensitivity analyses to show that any observed heterogeneity between family and hired labour is independent of the specific values of the ratio (w_f / w_m) used to compute total family or total hired labour input through converting female labour workdays into equivalent male workdays.

The bullock workdays variable is also distinguished by family source and hired source. The derivation used for the total bullock input variable (B) is simply the sum of hired bullock days and family bullock days. Potential heterogeneity in the input of bullock power from family and hired sources is not likely to occur, and is also of no interest in this study.

Preparing a measure of the land input variable (A) for the production function estimation, however, required some detailed computations. From the MPHBS farm management data it is possible to compute several alternative indicators of the land input. Since the dependant variable in the production function estimation is aggregate farm output, the main distinction relevant to defining the land input is between net sown area (which is a measure of the physical size of the land area that is cultivated by the farm household) and gross harvested area (which takes into account the actual cropping cycle whereby more than one crop may be planted on a particular plot in the reference survey period). Empirical applications of production functions have used both types of specification of the land input variable. Where data are available, gross cultivated area seems a more appropriate measure of the annual flow of services received from the physical land endowment of the farm (Carter, 1984).

The problem with defining the land input in terms of gross cultivated area with the MPBHS data set is that the information about land quality (paddy land or upland, irrigated or unirrigated) is available only at the level of the physical land endowment and not in terms of the land area allocated to specific crops. Several alternative ways of incorporating land quality variables were tried. In the production function specifications using ordinary least squares estimation, additional land quality variables can be introduced in a ratio format – such as the ratio of irrigated land to total physical land, or the ratio of paddy land in total land – by assuming that these ratios carry over to the gross cultivated area as well. Alternatively, in the specifications based on non-linear estimation, an aggregate quality adjusted composite land input can be computed directly from the data on physical land types and the index of cropping intensity in the same way as the aggregate labour input is constructed as a composite of family and hired labour. (Section 6.4.2 in Chapter 6 provides more details on how the composite land input variable is defined).

5.3.2 Labour Supply Function

From Equation 4.11 the labour supply regressions are of the form

$$(5.2) \quad L_s = \ell (w^*(\theta^*), E^*(w^*), S, \Phi, T) + e$$
$$= \ell (w^*(\theta^*), \pi^*(w^*) + E, S, \Phi, T) + e$$

where L_s is a measure of the total labour supply in all productive activities.

The labour time of the individual family members of the farm household has four general uses:

i.   unpaid work on own farm cultivation, including time devoted to subsidiary productive activities, such as livestock rearing, home processing of farm output for sale;
ii. work as a hired labourer in other households' farming activities;
iii. off farm non-agricultural hired labour work (such as in construction);
iv. household subsistence activities, such as home processing for own consumption, time spent on collecting firewood and water, etc.

The definition of labour supply adopted for the regression analyses of Chapter 7 – based on the definition followed in the coding of the MPHBS data – includes items (i) to (iii) only.[19] In the setting of the Nepal *tarai* there is very little off farm non-agricultural work available for individuals who continue to reside on their farm. The bulk of the labour days of work reported in the MPBHS sample therefore consist of item (i) and (ii).

The labour supply regressions are estimated at the level of individual family members of the farm household in order to allow for the effects of variation in individual characteristics, such as age and educational levels. The sample of individuals is limited to immediate family members residing in the household who can be identified through their relationships to the household head.

The dependant variable is the number of days of productive work reported by each economically active family member over a six-monthly cropping cycle. In the MPHBS data on the number of days worked for each economically active family member is actually available on a month to month basis. The monthly data has been aggregated into two six-monthly subtotals to correspond to the two phases of the survey data collected in the farm management module. This procedure helps to establish a link between the labour supply workdays (in six-monthly subtotals) and the data on the average wage rates paid by farm households that report hiring in labour in the summer and winter cropping cycles. In this way the seasonal labour supply behaviour of an individual can be modelled since there will be some differences in the average wage rates reported in the two cropping cycles. It is of course quite simplistic to assume that the operation of the rural labour markets in the study areas gives rise to two distinct seasonal wage rates, each of which is constant over an arbitrary six-monthly survey cycle. Nonetheless, this method of aggregating the monthly labour supply data offers a way of accounting for some seasonal wage variation to compensate for the missing data on the actual seasonal wage rates received by individual household members.

As noted above, the main definitional treatment regarding the wage rate variable is the imputed equivalence of the off farm wage rate for family labour (w) and the wage rate at which labour is hired in by the sample households in a particular village (w^h). In effect, the assumption that $w = w^h$ is forced since w is not directly reported in the MPHBS data set.[20] Although such an imputation for the off farm wage rate for family labour could be a source of unspecified bias in the estimates of parameters of the labour supply function, such errors are unlikely to be serious in the specific setting of the Nepal *tarai* for the several reasons noted below.

Given the fragmented nature of rural labour markets in Nepal, the primary place for off farm employment of family labour will be as hired labourers in the larger farms in their own or nearby villages. Hence, it is expected that $w = w^h$ on average for a sample of farm households from a particular village with limited non-agricultural employment prospects.[21] In a similar setting to the Nepal *tarai*, Bardhan and Rudra (1981) report a very limited locus of off farm employment in agricultural operations for wage labourers in West Bengal. If the local village economies were completely closed, and crop production was the only economic activity undertaken, there would be an exact equivalence between the average wage reported as paid out and the average wage rate reported as received, even in the presence of seasonal variations in the local wage rates. In reality local village economies are not completely closed, and there are other wage labour activities apart form crop production. But the effects these have in creating a major gap between the wages paid out and wages received by the sample households in a particular village is still likely to be small.

Secondly, the analytical results in Section 3.5 of Chapter 3 indicate that w is the correct wage rate to use in the labour supply regression for farm household members only if $w > w^h/\theta^*$ consistently in the sample data.[22] But in such situations there would be complete specialization in labour allocation: family labour would be completely supplied on the off farm market at wage w, and farm cultivation would be done completely by hired labour paid a wage w^h. Since one rarely observes such a pattern of specialization, it is more usual to find $w < w^h/\theta^*$. When the latter condition holds, the effective wage rate to be used in the labour supply equation is defined in terms of w^h/θ for the subset of households who hire in some labour for their farm cultivation. In such cases the off farm market wage rate, w, need not be known. The verification of whether $w < w^h/\theta^*$ can be obtained from the observed labour allocation pattern alone. The individuals who receive a high off farm wage rate would not be specializing in own farm production work, and hence can be detected through the usual occupation codes reported in the MPHBS.

To minimize the discrepancy related to varying wage rates due to differences in the type of work performed, irrespective of the relationship between w and w^h/θ^*, the labour supply regressions in Chapter 7 will be limited to individuals who report their main occupation to be farm operators or agricultural labourers. For instance, the reported wage rate for hired labour in crop production would not apply to the labour/leisure equilibrium of a household member who may do some work on the family farm but whose main occupation was as a shop keeper or a village carpenter. In these other principal occupations, the marginal returns to labour are likely to be very different from the average wage paid to casual labour hired in for crop production since the former represent semi-skilled occupations. But the variation in individual level wages are unlikely to be important in village settings for persons who report farm cultivation or agricultural labour as their main occupation, when farming is carried out with traditional cultivation techniques with little mechanization. The main difference will be between male and female adult wage rates, a difference that is recorded in the MPHBS data. Finally, the usual distinction between gross and net wages in settings such as the Nepal *tarai* is not likely to be important. The fixed costs to finding work on the local market will be minimal and there are no personal income tax wedges to consider.

In summary, the imputation $w = w^h$, which is forced by the data structure in the absence of direct observations of w, is defensible in the proposed estimation set up and in the setting of the Nepal *tarai*. In the labour supply regression results of Chapter 7, w^h becomes the effective wage for any family member reporting workdays in the hired labour category for a particular cropping cycle. For individuals in autarchic households, or in big farm households that hired in labour, the effective wage rates will be based on w^h with the appropriate adjustment for θ^* if labour heterogeneity is indicated.

The second main variable required for the labour supply regressions is the measure of household non-labour income, including the imputed value of farm profits. From Eq. 4.7, the nominal household non-labour income (NLY), corresponding to the linearized budget constraint, is defined as:

$$(5.3) \quad \begin{aligned} NLY &= E^* = \pi^*(w^*) + E \\ &= Y^*_N - w^* F^* + E \\ &\cong Y_N - w^* F + E \end{aligned}$$

where π^* is the maximized value of farm profit using w^* to value family labour inputs; and E is truly exogenous non-labour endowment income.

The specific derivation of NLY used in the labour supply regressions is the approximation based on Y_N given by the third line of Equation 5.3. Y_N represents the actual (not the optimized) value of the total returns from farm cultivation net of all purchased inputs, but including the implicit return to family labour applied to own farm work. The approximation for farm profit based on Y_N is preferable because Y_N is observed, while Y^*_N has to be computed from the first-order conditions for a specific functional form of the production function and its parameter estimates.[23]

The NLY variable based on observed data is easily computed: Y_N and E can be computed from the income categories reported in the MPHBS, F is directly reported in the farm management module, and w^* is derived in the manner described in Section 3.5. The optimized value of NLY would in addition be based on the estimated parameters of the production function, from which the optimum levels of F and H, as well as other variable inputs, have to be computed. For methodological reasons (relating to the two step estimation strategy) it is desirable that the labour supply regressions be independent of all of the other estimated parameters of the production function, except for the set Θ which reflects the extent of heterogeneity between family and hired labour, and from which the unobservable w^* is derived. Such independence can then be utilized for an alternative verification of the labour heterogeneity results of the production function. This process partially compensates for the fact that the production function and labour supply regressions are not jointly estimated to derive Θ.[24]

The main income categories included in the E variable are household income from gifts and remittances, land rental payments and rent from other farm assets, such as bullocks and farm machinery. In order to avoid having NLY become zero for most landless households (for whom $Y_N = 0$ by definition) the adopted definition of E also includes the imputed rental value of the family residence which is reported as an income source for every household in the MPHBS data set.[25]

5.4 Summary

This chapter presented a brief overview of the setting of the Nepal *tarai* and of the MPHBS data that is used in the estimation of the farm household model in Chapters 6 and 7. Although the MPHBS data from other regions of Nepal was also available, the empirical analysis in this book is restricted to the *tarai* region sample because of the limited usage of hired labour in other regions. The clustered sample design of the MPBHS gives rise to significant variation in the independent variables, such as wage rates, to permit estimation of the farm household model with cross-sectional data from the *tarai* region only.

This chapter also discussed in some detail the procedures and definitions followed in computing the main variables required for the regression analyses of the production and labour supply functions. One key feature is that the production function is specified at the level of aggregate farm output mainly because data on land quality are not available at the individual crop level. The labour supply equations are specified for individual family members who are in specific agricultural occupations. Individual level wage rates are not reported in the MPHBS data. Instead the village level wage rates for hired labour derived from the farm management component of the survey are used as proxies for the off farm wage rate for family labour. It is argued that, in the specific setting of the *tarai* region of Nepal, there will be approximate equivalence between the wage rate reported as paid out by sample households that hire in labour in a specific village, and the wage rates received by individuals from that same village who report working off farm on the hired labour market.

Appendix Table 5A.1   Distribution of Sample Households by District and Land Size Strata

Operated Land Strata*	Sample Districts by Region					Total Sample Size
	Eastern (Morang)	Central (Mahottari)	Western (Rupandehi)	Mid-west. (Banke)	Far-west (Kailali)	
Large farms	19	27	27	12	18	103
Medium farms	26	31	31	16	19	123
Small farms	79	42	44	33	29	227
Marginal farms	82	107	57	23	12	281
All Cultivators	206	207	159	84	78	734
Landless	96	95	29	21	32	273
Total	302	302	188	105	110	1,007

*The operated land strata definitions adopted in the *tarai* region sample of the MPBHS are as follows: marginal farms (less than 1.03 hectares), small farms (up to 2.73 hectares), medium farms (up to 5.44 hectares), large farms (above 5.44 hectares). See NRB (1988: 3).
Note: The actual sample of households used in the regression analyses in Chapters 6 and 7 differ slightly from the strata-wise sample size denoted in this table due to additional adjustments in the sample size from data cleaning, as explained later in those chapters.

Notes

1   This information on the size distribution of land holdings in Nepal is compiled from the *1991/92 Nepal Agricultural Census* (HMG/N, CBS, 1993, Table 3).
2   See HMG/N, Ministry of Agriculture, *Costs of Production for Major Crops in Nepal* 1985/86. The data from this publication is contemporaneous with the MPHBS data.
3   For this reason the MPHBS was conducted by Nepal Rastra Bank in close coordination with other ministries and concerned government agencies of Nepal. Prior to the MPHBS the only existing study on income distribution and consumption in Nepal, based on a national representative survey, was a report prepared by Nepal's National Planning Commission in 1983 using the data from a household survey it had conducted in 1976, (HMG/N/NPC, 1983).
4   The other two *tarai* districts in the MPHBS sample were Siraha in the Eastern region and Bara in the Central region.
5   A Poverty Assessment report on Nepal was prepared by the World Bank based on the MPHBS data set (World Bank, 1990) and the Bank has remarked favourably on the quality of the data in this survey. Subsequently, the World Bank itself conducted a Living Standards Measurement Survey (LSMS) of Nepal in 1996. Access to the LSMS data was provided too late to be used in the empirical analyses of this study. In any case, the farm management component of the LSMS is not as detailed as the MPHBS.
6   The data on non-monetized transactions and other own-household accounts proved useful in deriving the non-labour endowment income, especially for the poorer households, which is required for the labour supply regressions. See Deaton (1997) for a review on the format and quality of household survey data for developing countries.
7   The procedures followed in the first step of the MPHBS sample design are not relevant for the purpose of this study which is based on an *ad hoc* sub-sample of the *tarai* districts selected for the MPHBS. See NRB (1988), pp. 26-31 for full details.
8   With regard to local administrative units in Nepal, each village *panchayat* was subdivided further into nine wards. In the MPHBS sample design each selected *panchayat* was further broken down into three clusters, each consisting of three adjoining wards. A random selection of 2 out these 3 clusters was made from each selected *panchayat* in the third stage of the stratified sample design of the MPHBS.
9   The village *panchayats* in Nepal were renamed as village development committees after the political changes of 1991, but their boundaries have remained the same.
10   For a fuller description of these and additional features of the sample design procedure adopted in the MPHBS see NRB (1988), pp. 26-31.
11   The regression estimates presented in Chapters 6 and 7 assign equal weight to each household or to each individual in the sample, without accounting for the specific features of the MPHBS survey design. To the extent that the full sample of 1,007 households is used, the equal weight procedure is quite defensible since the sample size within districts in the MPHBS was chosen to be proportional to the district population At the final stage of the stratified sample design of the MPHBS there is also a more or less equal representation of households from the five land size strata (NRB, 1988: 29). This property is maintained in the five district sub-sample chosen for this study since the entire sample within a district is used.
12   NRB (1988: 230) Table 2A. Other sources of household income are income from wages and salaries, rental income, non-agricultural enterprise income (i.e. household trade and cottage industry activities) and gifts and remittances. Within the MPHBS definitions of own-account agricultural enterprise income, apart from crop production

which accounts for 81.2% of this sub-total, the minor components are animal husbandry (16.1%), horticulture (1.4%) poultry (1%) and fisheries (0.2%).

13   The absence of quantity data on chemical fertilizers is not a serious shortcoming in the estimation of the production function. At the time of the MPHBS, fertilizer distribution in Nepal occurred through a single para-statal agency that charged a fixed price throughout Nepal. So the value data is a good proxy for quantity if one ignores differences in the nutrient content of different fertilizers.

14   The MPHBS has data on monthly income from off farm wage labour at the household level. While this could be used to derive an average earning per off farm workday at the household level, this computation is likely to be affected by major reporting errors. Another problem is that this information cannot be used to derive wage rates for male and female workers separately.

15   As noted in endnote 12 above, the role of horticulture, poultry and fisheries components in total agricultural enterprise income is negligible for the average *tarai* region rural household. While animal husbandry, which on average accounts for 16% of total agricultural enterprise income, is an important ancillary activity to crop production, there is almost no use of hired labour for this component. Also, it is difficult to correctly impute what annual animal husbandry production and income is when sales and slaughter of farm animals also represent a depletion of capital stock.

16   The MPHBS Report indicates that child labour is concentrated more on what it classifies as household subsistence activities - collecting drinking water and firewood, home processing of food crops, etc. (NRB, 1988: 141, Table I).

17   Labour exchange arrangements can be viewed as hired labour contracts where payment is received in-kind in terms of the other participants' labour input. Of course, the incentive structure and costs of monitoring the effort of exchange labour differ from that of purely casual hired labour. In some specific situations the exchange labour arrangements may be based on close-knit groups or extended family/clan identities and hence be more similar to family labour. Since exchange labour workdays in the MPHBS data for the *tarai* region are so limited, its classification ultimately is not important. (See Tables 6.1a and 6.1b in Chapter 6 for the amount of exchange labour reported in the selected sample from the *tarai* districts of the MPHBS.)

18   Benjamin (1992) presents regression results for total labour demand in Javanese rice farms, using SUSENAS data, which strongly support the linear aggregation of the male and female labour inputs.

19   When data exists on activities included in (*iv*), this category can be treated as much a part of labour supply as the unpaid farm work under (*i*). Under optimal labour allocation conditions, the marginal returns to labour in all activities are equalized. Jacoby (1993) specifies his labour supply variable to include category (*iv*) activities.

20   Actually what is required is w_{ijk} where the i indexes an individual in farm household j for the cropping season k. The wage rates received by individual family members who report working off farm in the hired labour market may vary according to their skill and nature of the specific tasks performed, even in a particular local labour market. Unfortunately this information is not recorded in the MPHBS data nor can proxies for it be computed from other parts of the survey questionnaire.

21   This will be particularly true for female family members whose off farm mobility is even more restricted. The wages female labourers receive will be the wages other households in the village pay to female farm workers in that village.

22   θ^* is the ratio of the marginal product of hired labour and family labour at the optimum labour input allocation for farm production. See the discussion in Chap. 3, Section 3.5.

23   It may appear that the observed Y_N will consistently under estimate the optimal Y^*_N since the latter is an extremum value, and therefore Y_N on average would equal Y^*_N only if *all* households achieved their profit maximizing optimum input levels. However, in the presence of an additive error term in the production function, as specified in Equation 5.1, there will be random errors in farm output, conditional on the actual input levels applied. Hence the observed Y_N can be greater or less than the optimal Y^*_N (which is computed without considering the error term). Therefore Y_N can on average be a consistent approximation for Y^*_N without resorting to the extreme assumption that all farm households achieve their profit maximizing input levels.

24   Chapter 7 provides further details on how the labour supply regression results can be used to choose between alternative model specifications that discriminate between labour heterogeneity and homogeneity, independently of the production function estimation results.

25   The non-labour income (E) categories are derived from Section 3.1 of the MPHBS household survey questionnaire Form *Ga* for the cash income components, and Section 5 of Form *Ga* for the income in kind categories that are received as gifts.

6 Production Function Estimation Results and Tests for Labour Heterogeneity

6.1 Introduction

This chapter presents the results of a production function based procedure that tests for heterogeneity between family and hired labour inputs in crop production in the *tarai* region of Nepal. The production function regression results also serve as the first step of the sequential estimation strategy outlined in Chapter 4 for estimating a farm household model that allows for labour heterogeneity. The parameters of the production function that describe the specific nature of the labour heterogeneity will be used to generate the unobservable variables for the labour supply regressions in the second step (Chapter 7). Unlike the conventional model with "generated regressors" where the first step is used only to derive the unobservable variables for the second step, the estimated production function parameters presented in this chapter are of independent interest. These parameters describe the production component of the farm household model, irrespective of the nature of the heterogeneity detected. They determine the set of factor demand elasticities and the elasticities of input substitution that reflect the production side decision making of the farm household.

The test for labour heterogeneity is embedded in a production function structure that uses alternative ways of aggregating family and hired labour into a composite labour input. Consistent aggregation of family and hired labour, however, pre-supposes that the underlying production function is separable in the labour inputs (Berndt and Christensen, 1974). Therefore it is necessary first to test for the separability of the two types of labour in the aggregate farm production function. If separability is indicated, it allows a direct way of testing for labour heterogeneity together with the estimation of the complete set of parameters of the production function. In addition, as discussed in Chapter 3, the separable input structure is advantageous for generating the unobservable shadow wage rates for family labour needed to estimate the labour supply equations in the second step.

The statistical inference in this chapter is based on the primal estimation of a farm level production function. The problem with direct estimation of a production function when the variable inputs could be endogenous is well known (Griliches, 1984). A conventional resolution of this problem is to interpret the production function relationship in the light of the Zellner, Kmenta and Dreze (1966) framework of expected profit maximization by the farm household, which makes the error term independent of the inputs. This framework is used to justify the primal estimation approach adopted in this study. The alternative of using a dual estimating procedure (based on a cost or profit function) is not feasible in this context because the shadow wage rate for family labour under a general form of labour heterogeneity is unobservable.[1]

The empirical results presented in this chapter are based on two different subsets of the MPBHS sample for the five *tarai* region districts. One data set consists only of those households that utilize both family and hired labour in crop production (Sample I). The second data set consists of all farm households that report any use of family labour (Sample II). Sample I is obviously a subset of Sample II. These two data sets are described in Section 6.2, together with a summary table of the main variables of interest for the production function regressions.

The empirical results based on Sample I are given in Section 6.3, using a translog specification for an aggregate production function with family and hired labour as two distinct inputs. The main results are on the tests for the separability of family and hired labour. The values of the various elasticities of substitution derived from the estimated parameters are also reported for this specification. After establishing that the production function is indeed separable in these two types of labour, in Section 6.4 alternative functional forms for aggregating the two labour categories into a composite labour input are tested. These tests are based on data from Sample II that also includes households where no hired labour is used. Section 6.5 presents the full estimation results for the preferred labour aggregation functional form using Sample II. Section 6.6 tests for the robustness of the labour heterogeneity result through a series of sensitivity analysis. The full set of the elasticities of substitution and factor demand elasticities based on the preferred composite labour function, estimated over Sample II, are derived and discussed in Section 6.7. The last section provides a summary.

6.2 Data Summary

The general issues related to the structure of the MPBHS data set and variable definitions adopted in this study were discussed in Chapter 5. The main points noted there (in Section 5.3.1) that are relevant to this chapter were the decision to specify an aggregate production function, and to add up male and female labour inputs assuming perfect substitutability with unequal productivity.[2] This section gives further details about the actual sample of households used for the production function estimations in this chapter and a summary description of the main variables of interest.

The coverage of the farm management component of the MPHBS data set is limited to households that operate some land for crop production. Ignoring the purely landless households in the sample of the five selected *tarai* districts, farm management data is available for 734 cultivator households. (See Appendix Table 5A.1 in Chapter 5).

As noted in Chapter 5, the farm management data set was initially processed independently of the rest of the MPHBS data. Sample household identification codes were used to link the farm management data set with the data tapes received from Nepal Rastra Bank for the other parts of the MPHBS. Due to some errors in the household identification codes a few households in the farm management data set could not be linked in this manner, and they have been discarded. The farm management records for several other households were also discarded due to missing data, or because the conversion factors for the local units of measurements recorded in the farm management questionnaires were not noted in the MPHBS codebook. The actual usable sample size in the farm management data set created at IIDS, after data cleaning and verifying household identification codes, was 713 households.

For the estimation work of this chapter, a final adjustment to the sample was made by discarding those households whose recorded land holding was less than 0.1 hectares, or those households which did not report any use of own family labour as inputs in crop production.[3] These extra adjustments gave a final sample of 679 households, which is Sample II – the data set used in the main empirical work of Sections 6.4 to 6.7.

For the tests of separability of family and hired labour reported in Section 6.3, the sample is limited to those households that utilize both family and hired labour. The separability test is ideally done with a data set where every sample point has non-zero values for both labour inputs.[4] The number of sample households that use both family and hired labour is 279.

This constitutes Sample I, derived as a subset of the 679 households in Sample II so that the other exclusionary conditions are identical.

A summary description for the main variables used in the production function regressions in these two data subsets is given in Tables 6.1a and 6.1b. As expected, Sample I (which is limited to households that use both family and hired labour in crop production) contains a higher share of big farm households. The average operated farm size in Sample I is 4.11 hectares compared to 2.67 hectares in Sample II. In Sample I, 54% of the households are from the medium and large farm size strata (as defined in the MPHBS sample design) while the corresponding proportion in Sample II is only 33%.[5] These percentages, however, show that a substantial proportion of the bigger farm households is completely reliant on family labour. On the other hand, almost half of the households using some hired labour for crop production come from the small and marginal farm size categories. Because family size and labour supply behaviour of individual family members differ, neither the incidence of hired labour use nor the ratio of hired labour in total labour input are related to operated farm size in a simple proportional manner.

Another important difference between Sample I and II is the variation in the gender composition within the family and hired labour categories. The average number of workdays of family labour is substantially larger in Sample I because of the larger farm size endowments, but the extra application of family labour is biased towards male family members. The average workdays of female family labour is almost the same in Sample I and II. A relatively higher percentage of female hired labourers makes up for the lesser application of female family labour on bigger farms. Hence the estimation results on labour heterogeneity based on Sample I is more likely to suffer from a confounding influence of the differences in the gender composition of family and hired labour workdays.[6]

The average wage rates for male and female hired labour are also slightly lower in Sample I than in Sample II. These are unweighted average wage rates that do not fully reflect the clustered sample design of the MPHBS; but it is reasonable that a lower wage rate be associated with a sample where there is a higher incidence of hired labour. In terms of the ratio of the wage rates for female and male hired labour, the two samples are similar. This ratio (*wr)* ranges from 0.64 to 1, with a mean of about 0.85 in both samples. Assuming that male and female labour are perfect substitutes as inputs in farm cultivation, this ratio implies that female hired labourers are, on average, 15% less productive than male labourers.

Table 6.1a Farm Management Data Summary: Sample I [a] (N = 279)

Variable	mean	std. dev.	min.	max.
Composite farm output (kg.)	8,112	7,037	194	50,986
Composite output price (Rs./kg.)	2.79	0.39	2.42	4.64
Labour input variables: (in days)				
Exchange labour workdays – female	0.5	3.5	0	44
Exchange labour workdays – male	1.0	5.1	0	50
Exchange labour days – total	1.5	8.0	0	94
Family labour workdays – female	92.7	116.2	25	858
Family labour workdays – male	233.5	208.4	32	1,530
Total family labour workdays [b] – female	93.2	116.3	25	858
Total family labour workdays [b] – male	234.5	208.2	32	1,530
Hired labour workdays – female	119.0	156.9	8	960
Hired labour workdays – male	184.5	228.0	13	1295
Percentage of households using hired labour	100			
% of households using female hired labour	88			
% of households using male hired labour	92			
Share of hired labour in total labour input (%)	42.2	31.1	21	86
Total female labour workdays (all sources)	212.3	187.7	37	994
Total male labour workdays (all sources)	419.0	339.2	52	2,345
Total labour input (both genders): unadjusted	631.3	485.5	95	2,988
Total labour input (both genders): adjusted [c]	589.6	456.2	83	2,846
Share of labour in the value of production (%)	28.1	9.2	17	63
Nominal female hired labour wage (Rs./day)	8.94	1.32	6	15
Nominal male hired labour wage (”)	9.61	1.79	6.43	15
Ratio of female to male wage rate (*wr*)	0.86	0.07	0.64	1
Land input variables: (in hectares)				
Total farm cultivated area	4.11	4.08	0.1	37.4
Irrigated paddy land area	1.90	2.55	0	17.0
Unirrigated paddy land area	1.73	2.46	0	13.6
Total upland farm area	0.48	2.23	0	32.6
Gross area harvested	6.10	5.15	0.17	37.5
Cropping intensity ratio	1.59	0.44	0.64	3.02
Total bullock input days (B)	123.5	110.2	16	980
Total real material inputs quantity index (M)	572.0	565.9	82	3,436
Share of households in big farm category (%)	54.1			
Years of schooling of household head	1.97	2.02	0	8
Household's labour market exposure:				
Percentage that are autarchic	0			
Percentage with off farm labour supply	7.8			

[a] Sample I consists only of households that use both family and hired labour as farm inputs.

[b] Total family workdays includes exchange labour days.

[c] Adjusted total labour input is male workdays + (*wr*) x female workdays, where *wr* is the ratio of the female hired labour to male hired labour wage rate.

Table 6.1b Farm Management Data Summary: Sample II [a] (N = 679)

Variable	mean	std. dev.	min.	max.
Composite farm output (kg.)	5,278	5,738	104	50,986
Composite output price (Rs./kg.)	2.83	0.42	2.42	4.64
Labour input variables: (in days)				
Exchange labour workdays – female	1.0	6.2	0	92
Exchange labour workdays – male	1.5	7.6	0	121
Exchange labour days – total	2.5	13.1	0	213
Family labour workdays – female	89.9	97.7	8	858
Family labour workdays – male	185.1	172.7	11	1,530
Total family labour workdays [b] – female	90.9	98.1	8	858
Total family labour workdays [b] – male	186.6	172.8	11	1,530
Hired labour workdays – female	52.9	120.0	0	960
Hired labour workdays – male	82.6	176.9	0	1,295
Percentage of households using hired labour	44			
% of households using female hired labour	41			
% of households using male hired labour	43			
Share of hired labour in total labour input (%)	32.2	61.2	0	86
Total female labour workdays (all sources)	143.8	151.6	8	994
Total male labour workdays (all sources)	269.1	278.7	11	2,345
Total labour input (both genders): unadjusted	412.9	402.8	21	2,988
Total labour input (both genders): adjusted [c]	380.7	378.0	19	2,846
Share of labour in the value of production (%)	29.1	11.0	7	82
Nominal female hired labour wage (Rs./day)	9.21	1.28	6	15
Nominal male hired labour wage (")	9.86	1.86	6	16.5
Ratio of female to male wage rate (*wr*)	0.85	0.07	0.64	1
Land input variables: (in hectares)				
Total farm cultivated area	2.67	3.21	0.1	37.4
Irrigated paddy land area	1.27	2.05	0	17.0
Unirrigated paddy land area	1.05	1.85	0	13.6
Total upland farm area	0.35	1.54	0	32.6
Gross area harvested	4.11	4.31	0.1	37.5
Cropping intensity ratio	1.64	0.43	0.64	3.05
Total bullock input days (B)	83.0	87.6	6	980
Total real material inputs quantity index (M)	347.1	445.2	40	3,436
Share of households in big farm category (%)	31.2			
Years of schooling of household head	1.33	1.75	0	8
Household's labour market exposure:				
Percentage that are autarchic	23			
Percentage with off farm labour supply	33			

[a] Sample II consists of households using some family labour input in their crop cultivation.

[b] Total family labour includes exchange labour days.

[c] Adjusted total labour input is male workdays + (*wr*) x female workdays, where *wr* is the ratio of the female hired labour to male hired labour wage rate.

6.3 Estimation and Inference for Farms Using Both Family and Hired Labour

6.3.1 Input Separability

For a production process utilizing n inputs, separability of inputs X_1 and X_2 from the other inputs implies that the marginal rates of substitution between X_1 and X_2 are independent of the levels of the other $n-2$ inputs. That is, in the production function given by Equation 6.1

(6.1) $Q = f(X_1, X_2, ...X_n)$

inputs X_i and X_j are separable from X_k if

(6.2) $$\frac{\partial \left\{ \dfrac{\partial f(x)/\partial x_i}{\partial f(x)/\partial x_j} \right\}}{\partial x_k} = 0$$

since $\dfrac{\partial f(x)/\partial x_i}{\partial f(x)/\partial x_j}$ = the marginal rate of technical substitution between

input i and j. Two alternative equivalent representations of Eq. 6.2 are

(6.3) $$\frac{\partial \ln(\partial f(x)/\partial x_i)}{\partial \ln x_k} = \frac{\partial \ln(\partial f(x)/\partial x_j)}{\partial \ln x_k}$$

i.e. the elasticity of the marginal product of x_i with respect to x_k equals the elasticity of the marginal product of x_j with x_k (Chambers 1989: 43); and

(6.4) $\sigma_{ik} = \sigma_{jk}$

where σ_{ik} is the Allen partial elasticity of substitution (AES) between input i and k; and similarly σ_{jk} is the AES between input j and k (Berndt and Christensen, 1973a).

The intuitive meaning of separability implied by the conditions specified in Equations 6.3 and 6.4 is clearly demonstrated by considering what happens to the isoquants in (i, j) space when the level of input k is changed. If the slope of the isoquant is not affected by the level of x_k, then inputs i and j are separable from input k. This situation is shown in Figure 6.1(a). Changes in the application of input k lead to a parallel shift in the isoquants in (i, j) space. If there is a rotation in the isoquant when levels of input k are changed, as in Figure 6.1(b), then clearly the marginal rate of substitution between inputs i and j are affected by the levels of input k. Hence x_i and x_j are not separable from x_k.

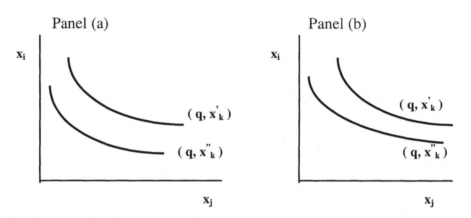

Figure 6.1 Input Separability

For the subset of the farm households that utilize some of both family and hired labour for crop cultivation, the separability of family and hired labour from the other inputs can be tested through parametric restrictions on the production function estimated with family and hired labour as separate inputs. The translog functional form is used for this purpose as an approximation to the true underlying production function (Denny and Fuss, 1977). The translog form allows for a flexible characterization of the relationship between family and hired labour.

The translog equation to be estimated is:

$$(6.5) \quad \ln Q = \alpha_0 + \sum_{i=1}^{5} \alpha_i \ln X_i + 1/2 \sum_{i=1}^{5} \sum_{j=1}^{5} \alpha_{ij} \ln X_i \ln X_j + \sum_k \delta_k Z_k$$

where family labour is indexed as input 1 (X_1), hired labour as input 2 (X_2), both measured in days; X_3 is gross harvested area, X_4 bullock days and X_5 material inputs. The elements of Z are additional explanatory variables that are not interacted with the elements of X. These include variables that control for the quality of land, such as the ratio of irrigated land and the ratio of paddy land in total holding size, and other variables such as the educational level of the household head. These are not interacted with the five main inputs to keep the number of estimated parameters manageable. Several dummy variables are also included in Z. The main ones are the regional dummies for the sample village clusters and a dummy for farm size.[7]

For the translog production function given by Equation 6.5, with family labour specified as X_1 and hired labour as X_2, following Denny and Fuss (1977), the conditions for various types of separability of the two types of labour from the other three inputs (X_3, X_4 and X_5) are given by the following restrictions on the parameters:[8]

weak separability (WS):

$$(6.6a) \quad \alpha_1 / \alpha_2 = \alpha_{13} / \alpha_{23} = \alpha_{14} / \alpha_{24} = \alpha_{15} / \alpha_{25}$$

partial strong separability (PSS):

$$(6.6b) \quad \alpha_{13} = \alpha_{23} = \alpha_{14} = \alpha_{24} = \alpha_{15} = \alpha_{25} = 0$$

complete strong separability (CSS) – Cobb-Douglas form:

$$(6.6c) \quad \alpha_{ij} = 0 \quad \text{for all i, j} \quad (i = 1 \text{ to } 5; j = 1 \text{ to } 5).$$

From the above restrictions it is clear that strong separability implies weak separability but not *vice versa*.[9] Finally, the complete strong separability restriction when imposed on a translog equation is not a restriction about separability alone but also of a specific functional form – the Cobb-Douglas.[10]

6.3.2 Tests for Labour Separability

The results of the statistical tests for the various forms of the separability restriction are given in Table 6.2. These results are based on an ordinary least squares (OLS) estimation of the translog production function of Equation 6.5. Two different test procedures are followed. In Test A, Eq. 6.5 is estimated without any additional restrictions (apart from the conventional one of symmetry between α_{ij} and α_{ji}). The restrictions implied by the various types of separability outlined above are then evaluated in separate independent tests, using the Wald Chi-square test statistic based on the unrestricted parameters. [11]

Test B follows a nested testing sequence where the subsequent tests are based on imposing the prior null hypotheses that have *not* been rejected. At the first level is a test for constant returns to scale (CRTS) of the five-input translog (TL) production function specified in Eq. 6.5. This restriction is not rejected. At the second level, Eq. 6.5 is re-estimated by imposing the CRTS restrictions, and the test for weak separability is based on the parameters of the CRTS restricted TL equation. If weak separability (WS) is not rejected at the second level, Eq. 6.5 is re-estimated with both the CRTS and WS restrictions imposed. Then, at the third level, the restrictions for partial strong separability are tested. If this is also not rejected, Eq. 6.5 is re-estimated with CRTS and PSS restrictions imposed, and a fourth level test for complete strong separability is implemented. The test sequence stops whenever a prior higher level null is rejected.

For nested tests the appropriate significance level of a specific test at the second and further levels depend on the significance levels of all the prior hypotheses in the sequence that were not rejected. Let $H_{0 \cdot i}$ be the i-th null hypothesis in a particular test sequence to be assigned a significance level of δ_i. This significance level is to be conditional on the sequential non-rejection of the i-1 prior hypotheses with assigned significance levels $\delta_1, \delta_2 \dots \delta_{i-1}$. Then the appropriate significance level δ_i for testing $H_{0 \cdot i}$, as specified in Denny and Fuss (1977), is

$$(6.7) \quad 1 - \prod_{j=1}^{i} (1 - \delta_j)$$

which is $= 1 - (1-\delta)^i$, if a common significance level (δ) is assigned to each sequential test level.

A significance level of 0.025 is chosen for each level in order to get reasonable significance levels in the second and third levels. With this choice, the overall test significance level is 0.040 at the second level, 0.073 at the third and 0.096 at the fourth level.

The test results reported in Table 6.2 are based on OLS estimation with White's correction for heteroskedasticity consistent standard errors. In the independent series of tests (Test A) only the WS null is not rejected. The other more restrictive forms of separability of hired and family labour are clearly rejected at the 5% significance level. The WS restriction (which has a Wald test statistic with a *p*-value of 0.45) is readily accepted by the data. In the separate test for constant returns to scale, which is independent of the separability restrictions, the null is also not rejected by the data.[12]

Table 6.2 Testing for Separability of the Labour Inputs

Test A: Independent separate tests based on base equation of unrestricted translog

test number	test for:	signif. level	Wald test statistic	df	*p*-value	Inference: reject null ?
i.	WS	.05	2.51	(3)	*0.473*	No
ii.	PSS	.05	17.01	(6)	*0.009*	Yes
iii.	CSS	.05	80.17	(15)	*0.000*	Yes
iv.	CRTS	.05	9.26	(6)	*0.159*	No

Test B: Sequential tests based on nested prior restrictions

test level	test for:	signif. level indiv. test	nested test	Wald test statistic	df	*p*-value	Inference: reject null ?
1.	CRTS	.025	.025	9.26	(6)	*0.159*	No
2.	given 1, WS	.025	.05	0.35	(3)	*0.950*	No
3a.	given 2, PSS	.025	.073	9.58	(3)	*0.002*	Yes
3b.[#]	given 2, CSS	.025	.073	59.6	(7)	*0.000*	Yes

[#] *Note*: The fourth level nested test sequence for CSS is not reached because PSS is rejected in the third step. Test level 3b is an alternative third level direct test for CSS based on the restrictions not rejected at levels 1 and 2 only (CRTS and WS).

CRTS = constant returns to scale   WS = weak separability
PSS = partial strong separability   CSS = complete strong separability
                                                                 (equivalent to Cobb-Douglas)
df   = degrees of freedom for the Wald Chi-square test statistic.
p-value = probability value of the computed Wald test statistic.[13]

In Test B the test sequence begins with the CRTS restrictions since constant returns to scale is a desirable property to have in production function estimation.[14] Since test number (iv) in the Test A series did not reject the null of CRTS, this condition is imposed at the first level in the Test B series. (Test A(iv) and B(1) are identical). After imposing CRTS in the Test B series, the second level test is for WS using the CRTS restricted TL parameters. Again WS is not rejected. At the third level, the estimated TL equation imposes both CRTS and WS *a priori*, and the test is for the additional restrictions of PSS. The results in Table 6.2 show this restriction is barely rejected (test level 3a), at the nested significance level of 7.3 %. The nested testing sequence properly terminates at this level. For completeness, an alternative third level test is also reported for complete strong separability – the Cobb-Douglas form – given prior restrictions of CRTS and WS. The Cobb-Douglas functional form is strongly rejected, with a Wald test statistic of 62.8 with 7 degrees of freedom (test 3b). The result of the nested test sequence is that the most specific restriction accepted by the data is constant returns to scale and weak separability of family and hired labour.[15]

In summary, both Tests A and B show that family and hired labour are only weakly separable from the other three main inputs (land, bullock days and material inputs). A more restrictive form of partial strong labour separability is rejected. The most restrictive form – complete strong separability, which implies a Cobb-Douglas equation – is strongly rejected.

6.3.3 Translog Parameter Estimates

The full regression results for the CRTS and WS restricted translog production function for aggregate farm output with family and hired labour as separate inputs are given in Table 6.3.[16] Three alternative sets of parameter estimates for the TL equation are presented. The specification in Model 1 imposes only the CRTS and WS restrictions discussed above. A potential problem with the parameter estimates of a translog equation, as with other flexible functional forms, is that the theoretical curvature conditions on the concavity of production functions could be violated (Diewert and Wales, 1987). Since the parameter estimates will be used to derive the elasticities of substitution and factor demands, which are meaningless if the estimated production function is not concave, it is necessary first to verify that the translog specification of Model 1 satisfies the concavity property at least at the mean of the sample data. It, unfortunately, does not.[17] Hence the CRTS-WS restricted specification of

Model 1 is re-estimated again with additional restrictions imposed to satisfy local and global concavity conditions.[18] The results are reported as Models 2 and 3, respectively, in Table 6.3.

The parametric estimates of the CRTS-WS translog equation, with local concavity imposed at the geometric mean (Model 2), are similar to the unrestricted estimates (Model 1). The log likelihood value decreases very marginally to 63.78 from 64.26 with the imposition of local concavity; and there are only minor changes in the estimates of the parameters of the interacted (second-order) variables. Local concavity restrictions change the sign of only three of the fifteen second-order coefficients, and all three are for parameters (α_{15}, α_{25}, α_{35}) that are not significantly different from zero in the unrestricted model. In both Models 1 and 2 the α_{12} parameter which represents the interaction between family and hired labour is significantly negative. This indicates that increasing application of the other labour input reduces the marginal product of each type of labour input, everything else held constant.

The global concavity restrictions (Model 3) however are too severe. They change the model fit and parameter estimates radically. The log likelihood value for Model 3 decreases to 32.85. All of the second-order coefficients (α_{ij}) become insignificant, which reduces this model to a Cobb-Douglas specification. But even as a Cobb-Douglas function the estimated elasticity parameters (α_i) are very unreasonable. They imply that the combined share of family and hired labour is less than eight percent of the value of output ($\alpha_1 + \alpha_2 = 0.076$), while the share of land is more than three fourths ($\alpha_3 = 0.765$).[19] The values of these share parameters are very reasonable in Models 1 and 2. The combined share of labour is about 37% and land is 43%, while the share of bullock power and material inputs are around 10% each. Model 3 illustrates the empirical problems likely to arise (as intimated in Chapter 3) when family and hired labour are treated as distinct inputs, if there is a high degree of substitution between them.

6.3.4 *Elasticities of Substitution*

The elasticity of substitution between family and hired labour can be computed from the parameter estimates of the locally concave, CRTS and WS restricted TL equation in Table 6.3 (Model 2). In a multi-factor setting there are many definitions of the elasticity of substitution between any two factors. The Allen partial elasticity of substitution (AES) denotes whether two factors are *p*-substitutes or *p*-complements (where *p* signifies price).

Table 6.3 Translog Function Parameters with Family and Hired Labour as Independent Inputs

Dependent Variable:	Log composite farm output quantity
Sample subset:	Sample I (N = 279)
Prior Restrictions:	Constant returns to scale and weak labour separability
Estimation Method:	OLS (Model 1); Non-linear least squares (Models 2 & 3)

		Model 1		Model 2		Model 3	
Additional restrictions:		none		local concavity		global concavity	
Variable		*coeff.*	*t- ratio*	*coeff.*	*t- ratio*	*coeff.*	*t- ratio*
Family labour (F)	$\alpha1$	0.223	7.28	0.227	8.10	0.043	2.95
Hired labour (H)	$\alpha2$	0.141	5.62	0.142	5.90	0.033	1.56
Land (A)	$\alpha3$	0.447	6.49	0.430	7.30	0.765	15.50
Bullocks (B)	$\alpha4$	0.104	3.76	0.104	3.73	0.092	3.45
Material inputs (M)	$\alpha5$	0.083	2.34	0.095	3.47	0.067	2.10
Education of Head		0.012	1.87	0.013	1.86	0.094	1.31
Big Farm dummy		0.041	1.39	0.036	1.23	-0.020	0.65
Second-order coefficients (x 10)							
FxF	$\alpha11$	0.437	3.87	0.442	3.93	-0.0002	0.03
HxH	$\alpha22$	0.581	6.45	0.590	6.81	-0.0069	0.08
FxH	$\alpha12$	-0.821	4.10	-0.838	3.80	0.001	0.08
AxA	$\alpha33$	-3.578	2.39	-3.690	2.49	-0.081	0.11
BxB	$\alpha44$	0.508	1.14	0.448	1.25	-0.016	0.11
MxM	$\alpha55$	1.114	1.74	0.657	2.10	-0.053	0.13
FxA	$\alpha13$	1.387	3.08	1.420	3.15	0.004	0.05
HxA	$\alpha23$	0.870	2.81	0.890	2.86	-0.024	0.02
FxB	$\alpha14$	-0.640	2.40	-0.659	2.55	-0.002	0.06
HxB	$\alpha24$	-0.402	2.36	-0.410	2.50	0.011	0.12
FxM	$\alpha15$	-0.364	1.45	-0.370	1.54	-0.003	0.06
HxM	$\alpha25$	-0.229	1.47	-0.232	2.47	0.019	0.11
AxB	$\alpha34$	1.188	1.65	1.030	1.52	0.036	0.11
AxM	$\alpha35$	0.133	0.20	0.350	0.69	0.065	0.12
BxM	$\alpha45$	-0.654	1.62	-0.400	1.67	-0.029	0.13

	Model 1	Model 2	Model 3
Adjusted R-Square	0.955	0.955	0.944
Residual sum of squares	10.31	10.34	12.91
Log Likelihood	64.26	63.78	32.85
Test for Cobb-Douglas form:	*(reject)*	*(reject)*	*(accept)*
Wald χ^2 test statistic (7 df)	59.60	86.90	0.19

A positive (negative) AES_{ij} signifies that inputs i and j are p-substitutes (complements), implying that the demand for factor i increases (decreases) when the price of factor j increases, holding output constant.[20]

In a general production function with n distinct inputs as in Equation 6.1, the Allen partial elasticities of substitution are given by:

$$(6.8a) \quad \sigma_{ij} = \frac{\sum_{k=1}^{n} X_k f_k}{X_i X_j} \cdot \frac{C_{ji}}{D}$$

where f_k is the partial derivative of $f(.)$ with respect to input X_k, D is the determinant of the $(n+1)$ by $(n+1)$ bordered Hessian matrix, and C_{ji} is the cofactor associated with the element f_{ji} in the bordered Hessian (Chambers, 1988: 33).

Equation 6.8a shows that the values of the Allen partial elasticities of substitution depend in a complicated way on the estimated parameters of the production function which are reflected in the bordered Hessian matrix.

The Hicksian elasticity of complementarity (HEC) between inputs i and j is defined only in terms of the first and second partial derivatives of the production function with respect to inputs i and j, as follows:[21]

$$(6.8b) \quad HEC_{ij} = \frac{Q \cdot f_{ij}}{f_i f_j} .$$

Following Hicks (1970), factors i and j are classified as q-complements or q-substitutes (where q signifies quantity) according to whether HEC_{ij} is positive or negative. Given non-negative marginal products, the sign of HEC_{ij} follows the sign of f_{ij} – the cross partial derivative of the production function with respect to the two relevant inputs. The values of f_{ij} measures the effect on the marginal product of input i (f_i) due to an exogenous change in the quantity of input j. Output is not held constant in the computation of HEC (as it is in the derivation of the AES). In fact, the HEC measures the degree to which two factors jointly contribute to a change in output (Sato and Koizumi, 1973).

If HEC_{ij} is positive, an increase in the quantity of factor j leads to an increase in the marginal product of input i, and so more of input i is also used, if input prices are held constant. In this case the increased use of input i and j jointly contribute to increase output. When HEC is negative, the increased usage of input j leads to a reduced demand for input i, and the combined effect on output depends on the relative magnitudes.

The direct elasticity of substitution (DES) is a multi-factor generalization of Joan Robinson's original interpretation of the elasticity of substitution as a measure of the response of the optimal factor input ratio of two inputs to a change in their factor price ratio (Chambers, 1988: 33).

$$(6.8c) \quad DES_{ij} \;=\; \frac{\partial \ln (X_i / X_j)}{\partial \ln (f_j / f_i)}$$

The DES is a short-run measure of the substitution between two inputs, holding output and all other inputs constant (McFadden, 1978: 80). It ranges between zero and plus infinity, and larger values signify easier substitution between the two inputs.

The values of the various elasticities of substitution between family and hired labour, computed from the parameter estimates of Model 2 in Table 6.3 at the geometric mean of the sample data, are as follows:[22]

AES between F and H	=	10.68
HEC "	=	-1.59
DES "	=	4.62

The positive value of AES indicates that family and hired labour are price-substitutes: as the price of one type of labour increases the quantity of the other type is increased along an isoquant. The estimated value of the AES is quite high, suggesting relative ease of substitution of one type of labour with another.[23] The negative HEC between family and hired labour indicates that increased usage of one type of labour reduces the marginal product of, and hence demand for, the other. This result, together with the positive AES, indicates that family and hired labour are clearly substitutes for each other in both the price and quantity sense. The computed value of the DES between family and hired labour, which is greater than 4.6, also indicates relative ease of substitution. In a similar computation of the DES between family and hired labour in Indonesia, the highest estimate of the DES was 1.68 (across different regions of Indonesia), with many being under 1 (Squires and Tabor, 1994).

The complete set of the Hicksian elasticities of complementarity for all five main production inputs at the geometric mean of the sample data is reported in Table 6.4. The imposition of local concavity at the geometric means insures that the matrix of the HEC_{ij} is negative semi-definite with all of the diagonal elements (own HEC's) being negative. (The imposition of weak separability of the two labour inputs, F and H, from the other three inputs also means that $HEC_{Fk} = HEC_{Hk}$, where $k = L$, B and M).

Table 6.4 Hicksian Elasticities of Complementarity
(with family and hired labour as separate inputs)

	Family lab.	Hired lab.	Land	Bullocks	Materials
Family labour	-2.54	-1.59	2.46	-1.77	-0.70
Hired labour		-1.77	2.46	-1.77	-0.70
Land			-3.32	3.28	1.85
Bullocks				-4.46	-3.37
Materials					-2.26

Source: Derived from the parameter estimates of Model 2 in Table 6.3 and evaluated at the geometric mean of the sample data.

The cross HEC's reported in Table 6.4 appear quite reasonable. Operated land area is a *q*-complement (positive HEC) for all other inputs. An increase in the land area increases the marginal product of all other inputs. Labour and bullock power are *q*-substitutes – increased application of labour reduces the marginal product of bullock power, which is a reasonable result. The least intuitive result is that labour and material inputs are *q*-substitutes. This would have been reasonable if material inputs included a lot of mechanized power inputs. But as noted in Chapter 3, the main elements of M are the value of seed and fertilizers.

6.3.5 Estimated Marginal Products

The regression parameters of Model 2 (Table 6.3) can also be used to derive the marginal products of the main inputs. These are presented in Table 6.5, estimated at the geometric mean of the data in Sample I. There is a strong indication of labour heterogeneity in these results, which show a substantially higher marginal product of family labour per day (Rs. 15.70) compared to the marginal product of hired labour (Rs. 10.61). The estimate of the marginal product of hired labour at the geometric mean is close to the average wage rate for hired labour paid out by the sample households. The difference is not statistically significant at the 5% level, given the reported standard error on the estimated marginal product of hired labour. On the other hand, the ratio of the marginal product of hired labour to family labour ($\theta*$ in the notation of Chapter 3) is 0.676. This estimate is significantly less than 1, even at the 1% level of significance.[24]

The estimates of the marginal products for the other inputs are also quite plausible. The marginal product of land is expressed as per hectare of gross cultivated area. This is the return to an additional hectare of multiple cropping, which is not the same as the return to increasing the operated land of the farm household by a hectare. The estimated marginal product for a bullock pair is higher than the marginal product of hired labour, which is consistent with the observed market wage rates for these inputs.

Table 6.5 Estimated Marginal Products of Inputs

Inputs	units	Marginal Product	
		value	stand. error
Family labour (F)	Rs. per day	15.70	1.93
Hired labour (H)	Rs. per day	10.61	1.81
Land (A)	Rs. per ha.	2,371	325
Bullocks (B)	Rs. per day	19.10	5.12
Material inputs (M)	per unit of real expenditure	4.35	1.25
Memorandum items:			
Average daily wage (male hired labour, Sample I)		Rs. 9.61	
Ratio of the marginal product of hired labour to family labour		# 0.676	0.083
		# significantly less than 1	

Source: Value of marginal products computed at the geometric mean of the data using the regression parameters of Model 2 in Table 6.3.

6.3.6 Alternative Estimates with a Homogeneous Labour Aggregate

The translog equation with family and hired labour as separate inputs gives reasonable estimates for the production function parameters as well as the marginal products of the inputs, and the various elasticities of substitution. The reasonableness of these results, however, needs to be checked against alternative specification that may be preferred to the one that has F and H as distinct inputs. Table 6.6 presents the parameter estimates for a TL specification that assumes that family and hired labour are homogeneous inputs. In this specification (Model 4) there are only four main inputs which are interacted with each other: total labour (L, which is simply the sum of F and H), land, bullock days and material inputs. Constant returns to scale and local concavity are imposed as in Table 6.3.

A comparison of the parameter estimates between Model 2 (Table 6.3) and Model 4 (Table 6.6) reveals only minor changes in the first-order coefficients (the α_i's). The coefficient α_1 on total labour in Model 4 is almost an exact sum of the corresponding coefficients on family and hired labour in Model 2. But there are bigger changes in some of the second-order coefficients (α_{ij}'s). The coefficient on the total labour quadratic term (L x L) changes sign in Model 4, and there is a large change in the value of the coefficient on the land area quadratic term (A x A).

The effect of these changes in the estimated parameters on the underlying household behaviour can be illustrated by computing the values of the Hicksian elasticities of complementarity for Model 4. These are reported in Table 6.7 (computed at the geometric mean of Sample I). Comparing the values of the HEC's in Table 6.7 and Table 6.4 shows that the input relationships based on a model with aggregate homogeneous labour does differ considerably from the relationships based on treating family and hired labour as separate inputs.

Table 6.6 also reports on a test for common production function parameters in the sample of farms that use both family and hired labour (Sample I) and farms that use only family labour.[25] There are no *a priori* reasons to believe that the production technology available to households that use hired labour is different from households using only family labour. This proposition is tested with the homogeneous labour aggregate because with this specification the model can be estimated by ordinary least squares for which the F test for the stability of parameters in different samples is exact.[26] The total number of parameters estimated with the homogeneous labour aggregate is 47 (including 31 sample village cluster dummies). The computed F test statistic is less than the critical value at the 5% significance level. Hence, the null of a common production technology for all farm households, irrespective of whether they use hired labour or not, is not rejected.

Returning to the parameter estimates relating to Sample I households only, Model 4 in Table 6.6 and Model 2 in Table 6.3 are not nested within each other. Therefore the diagnostics tests on which specification is

Table 6.6 Translog Production Function Parameters with Homogeneous Labour

Dependent Variable:	Log real aggregate farm output		
Sample subset:	Sample I - households using family and hired labour (N = 279)		
Prior Restrictions:	Constant returns to scale		
Estimation Method:	OLS with heteroskedasticity consistent errors		
		Model 4	
Additional restrictions :		local concavity	

Variable		*coefficient*	*t- ratio*
Total labour (L)	α1	0.367	5.78
Land (A)	α3	0.449	6.26
Bullocks (B)	α4	0.105	3.87
Mat. Inputs (M)	α5	0.079	2.75
Education of Head		0.01	1.65
Big Farm dummy		0.024	0.9
Second order coefficients (x 10)			
L x L	α11	-1.35	0.64
A x A	α33	-6.08	2.51
B x B	α44	0.37	0.95
M x M	α55	0.66	1.07
L x A	α13	3.95	1.97
L x B	α14	-1.83	2.13
L x M	α15	-0.77	1.00
A x B	α34	1.74	1.74
A x M	α35	0.38	0.497
B x M	α45	-0.27	0.73

Adjusted R- Square	0.955
Residual sum of squares	10.22
Log Likelihood	65.4
Test for Cobb-Douglas form:	
Wald test statistic (Chi-Square with 6 df)	12.27 *reject*
# Test for common production technology:	
F test statistic (with 48 and 583 df)	1.36 *don't reject*

This is a test for common parameters of the production function in the sample of farms that use both family and hired labour (Sample I) and farms that use only family labour. The latter subset consists of farms that are in Sample II but not in Sample I. The sample size is 400 (= 679 - 279) for this subset of households which use only family labour.

**Table 6.7 Hicksian Elasticities of Complementarity
(with homogeneous aggregate labour)**

	Total labour	Land	Bullocks	Materials
Total labour	-2.73	3.40	-3.76	-1.67
Land		-4.24	4.68	2.08
Bullocks			-5.17	-2.31
Materials				-1.02

Source: Computed at the geometric mean of the data using the estimated regression parameters of Model 4 in Table 6.6.

preferred are not straightforward. As an illustration it is interesting to note that the standard goodness of fit criteria for non-linear regressions, such as the log likelihood and the residual sum of squares, indicate that Model 4 with the homogeneous labour aggregate provides a better fit for the sample data. This is a surprising result given that Model 2 is a flexible functional form which provides for a general form of inter-relationship between family and hired labour. If family and hired labour were truly homogeneous inputs, Model 2 could have reflected such a relationship instead of the estimated wide divergence in the marginal products of the two labour inputs at the mean of the data.

It appears that production functions that treat family and hired labour as distinct inputs could be mis-specified when in fact the two types of labour are very close substitutes. This was clearly evident in the Cobb-Douglas specification with family and hired labour as distinct inputs – the parameter estimates in Model 3 (Table 6.3) were nonsensical. Similar problems may carry over to more general functional forms as well, given that Model 4 with a homogeneous labour aggregate performs better than a flexible specification with family and hired labour as distinct inputs. The preferred alternative is to look at ways of combining the family and hired labour input categories into a composite labour input aggregate without *a priori* imposing homogeneity. This alternative approach is taken up in the next section of this chapter.

6.4 Testing for Alternative Aggregates of Family and Hired Labour

The separability results of Section 6.3.2 imply that family and hired can be consistently aggregated into a single composite labour input. Specifying the production function with a composite labour input also means that the parameter estimates can be based on Sample II, which includes households that do not report any use of hired labour for crop production. Such households constitute almost 60% of the total available sample of 679 land cultivating farm households created from the MPBHS *tarai* region sample. The production technology available to households that utilize some hired labour is unlikely to be different from that available for households using only family labour; and the *F* test reported in Table 6.6 did not reject this null hypothesis. It is appropriate, therefore, to estimate a common production technology for all households, specified on the basis of a composite labour input, irrespective of whether or not a particular household reports any use of hired labour.

An alternative specification of the translog production function that still maintains family and hired labour as distinct inputs when estimated over Sample II is inappropriate. In this case a large proportion of households would have a zero level of input of hired labour. *Ad hoc* procedures of re-scaling the data by converting the zero values to one (or ten) so that logarithmic values can de defined, while commonly used, are inappropriate for this data set because of the high proportion of zero values for hired labour.[27]

The customary approach to creating a composite variable from two or more separable inputs is to create a value-share weighted linear or log-linear (Tornquist) index. For an underlying translog production process, a Tornquist composite of separable inputs would be an exact index (Diewert, 1976). This procedure is not feasible here because zero values of hired labour are common, and also because the relevant "price" of family labour is not observed. Indeed the main purpose of carrying out the production function estimation is to decipher whether and how this (shadow) price of family labour differs from the observed market wage rate.

Consequently, alternative functional forms are specified for creating a composite quantity of family and hired labour that allows for zero values of hired labour. The composite labour input then becomes a single input in the main translog production function for aggregate farm output. The specification for the composite labour aggregator function is nested into the translog production function of Equation 6.5, which is now re-specified in terms of the four main inputs that are interacted: composite labour, land, bullocks and material inputs. The composite labour variable can be interpreted as the total input of labour in effective units (as opposed to standard time inputs of days or hours); and it can be referred to in short as effective labour.

6.4.1 *Effective Labour Functions*

Five alternative functional forms are chosen to create an effective labour (Le) composite from the observed levels of family (F) and hired labour (H) inputs. (These functional forms denote the g (.) function in the terminology of Chapter 3, and as specified in Equation 4.9). The selected forms allow for different possibilities of efficiency differences as well as for a constant or a varying elasticity of substitution between family and hired labour in the "production" of effective labour. The specific functional forms chosen, which are adapted from Deolalikar and Vijverberg (1983), are:

6.9. C1 Homogeneous labour: $\quad Le = F + H$

6.9. C2 Linear heterogeneous: $\quad Le = F + \theta*H \qquad\qquad\qquad \theta > 0$

6.9. C3 CES:[28] $\qquad\qquad\qquad Le = [F^{-\rho} + \theta*H^{-\rho}]^{-1/\rho} \quad \theta > 0, \ \rho \geq -1$

6.9. C4 Generalized Linear (GL): $Le = F + \theta*H + 2*\delta*(H*F)^{\frac{1}{2}} \quad \theta > 0$

6.9. C5 Ratio: $\qquad\qquad\qquad Le = (F + H) * (F / (F + H))^{\mu} \quad |\mu| \leq 1$

Once separability is established for a subset of inputs, there are two additional properties of the $g(.)$ aggregator function that determine whether two different production inputs are homogeneous or heterogeneous:

i.   whether $g(.)$ has a linear form such that the input components are perfect substitutes for each other, implying an infinitely large AES;
ii. whether the ratio of marginal products (the marginal rate of technical substitution between input pairs) is equal to one.

The five functional forms specified above allow for different possibilities on points (i) and (ii). C1 and C2 imply an infinite elasticity of substitution between F and H. The CES composite implies a constant elasticity of substitution. The Generalized Linear and the Ratio form imply a varying elasticity of substitution that depends on the levels of F and H and the parameters.

The CES functional form is normally specified with the restriction that $\rho \geq -1$ which is required for concavity to ensure the isoquants are well behaved. When the CES specification is a complete production function in itself (without being nested into another function) concavity implies that all the inputs be *p*-substitutes with positive AES (or equivalently *q*-complements with positive HECS). A necessary condition for concavity is that $\rho \geq -1$ because when $\rho < -1$ the AES between the inputs are negative, implying the inputs are *p*-complements.[29] But when a CES aggregator function such as C3 is nested within a higher level production function, such as the translog of Equation 6.5, the restriction $\rho \geq -1$ in the CES nest is not strictly required. The appropriate theoretical restriction is that the overall production function be concave. This concavity condition will also depend on the other parameters of the translog function, and hence can be satisfied even if $\rho < -1$ in the CES nest for composite labour.

In the empirical results given below the CES composite for family and hired labour is estimated both with and without the restriction that $\rho \geq -1$. The restriction on ρ has a central bearing on the nature of labour heterogeneity that is the subject of interest. When the CES composite is estimated with the restriction $\rho \geq -1$, it forces F and H to be *q*-complements with a positive HEC – i.e. an increase in the input level of F necessarily increases the marginal product of H, and *vice versa.*[30] But if F and H are to be *q*-substitutes, as is more likely the case, then the CES form for composite labour must allow for $\rho < -1$.

The Ratio composite is equivalent to Revankar's Variable Elasticity of Substitution (VES) production function, for which it can be shown that the AES between F and H (σ_{FH}) is equal to $1 + (F/H)*(1/\mu)$ (Revankar, 1971).[31] For $\mu = 0$, this form reduces to the homogeneous labour case (C.1) with an infinite elasticity of substitution. For non-zero μ, the AES between hired and family labour is increasing or decreasing in the ratio of F to H, depending on whether μ is positive or negative.[32] That the elasticity of substitution between F and H could be related to the ratio of F to H is consistent with a particular approach to modelling labour supervisory/monitoring costs on the premise that the hired labourer's work intensity or effort can be increased through more effective monitoring if family labour is also working alongside with hired labour (Frisvold, 1994).

For the GL functional form, the AES between F and H is given by:[33]

$$(6.10) \quad \sigma_{FH} = 1 + (F + \theta H)/Le + 2*(H*F)^{1/2}/(\delta*Le)$$

In this formulation, the elasticity of substitution depends in a more complicated way on the actual levels of F and H, and not just on their ratio. In Equation 6.10 σ_{FH} can be negative only if $\delta < 0$ (though this is only a necessary but not a sufficient condition). Hence, although the concavity of the GL form can be violated when $\delta < 0$ (Diewert, 1971), this restriction is not imposed in the estimations reported in this chapter since the overall translog production function can still be concave with $\delta < 0$.

For the functional forms specified in the set of Equation 6.9 above, it is clear that with appropriate parametric restrictions on ρ and δ, both the CES and the GL forms reduce to the linear heterogeneous composite (C2); and the latter obviously reduces to the homogeneous composite (C1) when $\theta = 1$. Hence, several combinations of the first four specifications can be tested in the framework of nested models. The ratio composite (C5) is unrelated to the GL, CES or linear heterogeneous forms; it reduces only to the homogeneous case.

6.4.2 *Effective Land Function*

In addition to the composite function for effective labour, the land input variable is also specified as a separable nest, allowing for heterogeneity between different types of land and also allowing for variation in land use due to differences in the multiple cropping intensities of farms. The specification chosen for the effective unit of the land input (Ae) is:

$$(6.11) \quad Ae = A_c \cdot CRINT^{\kappa}$$

where A_c is the physical cultivated area of the farm or the land endowment available for crop production, already measured in effective units; and CRINT is the cropping intensity ratio which is equal to A_g / A_c, where A_g is the gross area harvested of different crops, which is the sum of the land area allocated to each crop grown on the farm.

This specification treats the increases in the physical area of the farm differently from increases in the harvested area due to more intensive multiple cropping. When $\kappa = 1$, Equation 6.11 is just an identity with effective land being the same as gross harvested area (A_g). If as expected $\kappa < 1$, the returns from increased multiple cropping are less than the returns from increasing the physical land area of the farm. This is consistent with decreasing returns to multiple cropping given a fixed land area.[34]

The effective cultivated land area (A_c) available to the farm is itself aggregated from three different land categories, assuming a linear form. This assumes perfect substitutability but allows for differing unit productivity among the three land input categories just as in the case of the labour inputs. The farm management data set of the MPHBS actually identifies four different types of land: irrigated paddy land (*khet*), irrigated upland (*pakho*), unirrigated *khet* and unirrigated *pakho*. In the final specification for the effective land input function the distinction between irrigated and unirrigated land in the upland category did not turn out to be important and is ignored.[35]

The final specification selected for A_c is

$$(6.12) \quad A_c = A_{c \cdot Pi} + \gamma_1 A_{c \cdot Pu} + \gamma_2 A_{c \cdot U}$$

where $A_{c \cdot Pi}$ is irrigated paddy land, $A_{c \cdot Pu}$ is unirrigated paddy land and $A_{c \cdot U}$ is total upland. This specification means γ_1 and γ_2 are conversion factors which measure A_c in units of total irrigated paddy land. The expected relationship is $< 0 < \gamma_2 < \gamma_1 < 1$ since even unirrigated paddy land is usually more productive than upland, whether or not the latter is irrigated.

Just as in the case of the Le function for effective labour, the Ae function for effective land input is nested in the overall translog production function of Eq. 6.5, which is now defined in terms of four main inputs.

The estimating equation is

$$(6.13) \quad \ln Q = \alpha_0 + \sum_{i=1}^{4} \alpha_i \ln X_i + 1/2 \sum_{i=1}^{4} \sum_{j=1}^{4} \alpha_{ij} \ln X_i \ln X_j + \sum_{k} \delta_k Z_k$$

where X_1 is effective labour, X_2 is effective land and X_3 and X_4 are bullock power and material inputs. As before Z is the set of other variables not interacted with the set of X.

6.4.3 Labour Heterogeneity Test Results

Equation 6.13 is estimated by non-linear least squares for each of the five effective labour composites (C1 to C5), and the common effective land composite nested into the main production function.[36] Table 6.8 presents the principal results of interests, focussing on the parameters of the effective labour function which determine the nature of labour heterogeneity, and on summary statistics for the fit of a specific model.

The main inference to be drawn from Table 6.8 is that among the set of nested models (C4 → C2 → C1, and C3 → C2 → C1), the restrictions implied by the general linear composite of Model C2 are accepted. But the additional restriction that makes family and hired labour homogeneous inputs (Model C1) is clearly rejected. In comparison to the simple linear composite (Le = F + θH), the more general form of a CES composite (C3) or a GL composite (C4) do not offer any additional explanatory power to the production function regressions.

In the CES nest with ρ unrestricted, the estimated value of ρ is -1. 03, but it is not significantly different from -1.[37] This reduces the CES function to a linear heterogeneous composite (C2). Similarly in the GL specification the δ parameter is not significantly different from zero; so the GL also reduces to the linear composite function.

At the next step, however, there is a significant difference between the linear composite and the homogeneous model (C1) since the θ parameter is significantly less than one. The estimated value of θ in Model C2 is 0.751, with a standard error of 0.107. The Wald Chi-square test statistic for the hypothesis that $\theta = 1$ is 5.44, with a *p*-value of 0.03. Hence the null of $\theta = 1$ is rejected at the 5% significance level. The same inference is obtained from the Likelihood Ratio test in comparing Models C1 and C2.[38]

In the Ratio functional form the estimate of the μ parameter is positive – which is consistent with family labour being more productive than hired labour.[39] But its estimated value of 0.042 is not significantly different from zero. Therefore the Ratio format can also be rejected. Although the Ratio and the linear composite forms (C5 & C2) are not nested, the latter is clearly preferable in terms of the log likelihood value.[40]

The final outcome is that the preferred effective labour aggregator function is C5 (Le = F + θH), where θ is significantly less than one. This implies that, although family and hired labour are perfect substitutes, there is a constant efficiency difference between the marginal product of family and hired labour, with family labour being more productive. It is worth noting that all specifications that allowed F and H to be *p*-complements with a negative AES (the GL with $\delta < 0$, the Ratio with $\mu < 0$, and the unrestricted CES with $\rho < -1$) were clearly rejected.[41]

The above result that family labour is more productive per unit of time is consistent with the empirical findings of Frisvold (1994) and Deolalikar and Vijverberg (1983) for India. However, their preferred aggregator functional forms are different. Deolalikar and Vijverberg (1983 and 1987) reject a linear composite and hence find evidence for imperfect substitutability between family and hired labour. Foster did not actually test alternative specifications of the Le function; he specified the Ratio format *a priorily* and found that μ was positive and significantly different from zero.[42] This also implies imperfect substitutability.

Table 6.8 Summary Results on Tests for Alternative Composite Labour Functions (using Sample II)

Model	Composite function	Parameters of interest	Estimated value	Stand. error	Restriction	Wald test statistic	p-value	Inference	^ R-Sq.	Log likelihood
C4	GL	θ	0.751	0.107	θ = 1	5.44	0.03	Reject	0.965	141.06
		δ	0.0075	0.087	δ = 0	0.007	0.93	Do not Reject		
C3	CES (ρ restricted)	θ	0.780	0.114	θ = 1	3.68	0.05	Reject	0.965	141.06
		ρ	-1 #	-	ρ = -1 #	-	-	# Restriction is imposed		
	CES (ρ unrestricted)	θ	0.757	0.115	θ = 1	4.46		Reject	0.967	141.07
		ρ	-1.029	0.144	ρ = -1	0.04	0.84	Do not Reject		
C5	Ratio	μ	0.042	0.034	μ = 0	1.52	0.23	Do not Reject	0.965	139.34
C2	Linear	θ	0.751	0.107	θ = 1	5.42	0.02	Reject	0.967	141.06
C1	Homogeneous	θ = 1							0.964	138.56

^ R-Square between observed and predicted values.

Inference on Nested Models: Restrictions on Model C4 and Model C3 that reduce them to Model C2 are not rejected. The restriction that reduces Model C2 to C1 is rejected.

6.5 Complete Results for the Linear Composite Labour Model

Table 6.9 gives the complete regression results for the linear effective labour composite (Model C2 version of Eq. 6.9). The only prior restriction imposed is constant returns to scale. The set of estimated parameters makes this specification concave at the geometric mean so additional concavity restrictions have not been imposed (as was done in Model 2 of Table 6.3).

The set of parameters of the TL production function given in Table 6.9 appears reasonable. At the geometric mean of the data the estimated input elasticities are 0.327 for effective labour, 0.487 for effective land, 0.076 for bullock power and 0.111 for material inputs. These are reasonable estimates of the share of these inputs in the value of farm output at the mean of the data. Several of the second order coefficients are statistically insignificant, but the joint test that the model reduces to a Cobb-Douglas specification is rejected. The estimates of the parameters of the land composite function are also as expected. The coefficient on the cropping intensity component of total land use (κ) is less than one. The γ_1 and γ_2 parameters indicate that unirrigated paddy land has about 92% of the productivity of irrigated paddy land, and upland has about 84% of the productivity of irrigated paddy land.[43] Each of the κ, γ_1 and γ_2 parameter is individually significantly less than one and the joint test for land homogeneity ($\kappa = \gamma_1 = \gamma_2 = 1$) is also rejected at the 5% significance level.

The main implication of the linearly heterogeneous labour composite is a constant difference in the marginal productivity of family and hired labour represented by θ. The implied values of the marginal products based on the regression parameters and computed at the geometric mean of the data are given in Table 6.10. For comparison, the marginal product of homogeneous labour (based on the composite function C1) is also reported.

From the estimates for the C1 labour composite, the marginal product of a homogenous labour aggregate, at the mean of the input variables, is Rs. 12.34 for an additional workday of a male labourer. This estimated marginal product is about 25% higher than the average male wage rate in the Sample II data (Rs. 9.86 per day). This is a relatively large discrepancy in a setting where the normal presumption would be for the marginal product of total labour to be lower than the wage rate paid out to hired labour. Apart from a general indication of surplus labour in the *tarai* region of Nepal, the specific wage rates derived from the MPBHS data are also likely to reflect peak-season hired labour wage rates rather than an annual average wage rate.[44]

Table 6.9 TL Production Function Parameters with Linear Heterogeneous Labour Composite

Dependent Variable:	Log real composite farm output			
Data subset:	Sample II (all households using some family labour)			
Prior Restrictions:	Constant returns to scale			
Estimation Method:	Non-linear least squares			

		Model C2 $Le = F + \theta H$		
Variable		*coefficient*	*asymptotic std. error*	*asymp. t- ratio*
Effective labour (Le)	$\alpha1$	0.327	0.044	7.45
Effective land (Ae)	$\alpha2$	0.487	0.051	9.59
Bullocks (B)	$\alpha3$	0.076	0.020	3.73
Material inputs (M)	$\alpha4$	0.111	0.022	5.09
Education of Head		0.007	0.005	1.37
Big Farm dummy		0.029	0.022	1.34
Second-order coefficients				
Le x Le	$\alpha11$	-0.051	0.105	-0.49
Ae x Ae	$\alpha22$	-0.21	0.147	-1.43
B x B	$\alpha33$	0.033	0.015	2.24
M x M	$\alpha44$	0.067	0.042	1.58
Le x Ae	$\alpha12$	0.146	0.110	1.33
Le x B	$\alpha13$	-0.077	0.043	-1.80
Le x M	$\alpha14$	-0.018	0.052	-0.34
Ae x B	$\alpha23$	0.079	0.047	1.67
Ae x M	$\alpha24$	-0.015	0.060	-0.25
B x M	$\alpha34$	-0.034	0.027	-1.26
Labour nest parameter	θ	0.751	0.107	0.107
Land nest parameters	γ_1	0.923	0.046	0.046
	γ_2	0.836	0.079	10.54
	κ	0.743	0.065	11.38

R-Sq. between observed & predicted:	0.97		Log Likelihood	141.06
Residual sum of squares	23.71			

Wald Chi-Square test statistics on restrictions:

labour homogeneity	$\theta = 1$	df =1	5.44 *		*reject*
land homogeneity	$\gamma_1 = \gamma_2 = \kappa = 1$	df =3	11.56 **		*reject*
Cobb-Douglas form		df =6	12.72 *		*reject*

Note: ** indicates null is rejected at the 1% significance level; * at the 5% level.

The model with the Le = F + θH specification helps to reconcile the observed gap between average labour productivity and wage rates. With this specification the estimate of the marginal productivity of hired labour at the mean of the data is only Rs. 9.19 per male workday, which is slightly lower than the average reported wage rate. This result is more in conformity with profit maximizing behaviour of the sample households. Whenever hired labour is used, the estimated marginal product on average is close to the market wage rate for hired labour.

The effect of the efficiency difference between family and hired labour is that it makes the effective return to family labour higher in own farm production as a substitute for hired labour. Family labour in such situations should act as if the effective wage rate for own farm work is higher than the market wage rate paid to hired labour. This should induce a greater labour supply response of family members in big farm households. Whether the actual labour supply behaviour of family members in households that hire in labour is consistent with the specific form of labour heterogeneity indicated in the production function estimation (with θ = 0.75) is taken up in Chapter 7.

Table 6.10 Estimated Marginal Products of Labour with Linear Composite

Model	Inputs	units	Marginal Product	
			estimate	stand. error
C1 (Le = F + H)	Labour (homogeneous)	Rs. per day	12.34	1.51
C2 (Le = F + θH)	Family labour (F)	Rs. per day	12.24	1.64
	Hired labour (H)	Rs. per day	9.19	2.12
Memo item :	Average wage rate of male hired labour	Rs. per day	9.86	1.86 #

Source: Value of labour marginal products computed at the geometric mean of the Sample II data, using estimated regression parameters of Models C1 and C2.
Standard deviation of male wage rate in Sample II data.

6.6 Sensitivity Analysis

Given the critical role that θ plays in the characterization of labour heterogeneity as estimated for the production side of the farm household model, and in the subsequent specification for the labour supply equations, it is important to establish the robustness of the result that θ is significantly less than one. Table 6.11 presents the results for a sensitivity analysis of the estimate of θ derived from alternative restrictions placed on the translog production function with the linear heterogeneous nest for effective labour. In all of the specifications reported in Table 6.11, the null hypothesis that $\theta = 1$ is rejected at the 5% significance level.

Table 6.11 Estimates of the θ Parameter with Alternative Functional Forms and Prior Restrictions

Model Le = F + θH nested within:	Restrictions	Estimate of θ	Stand. error	Testing Null $\theta = 1$		Log likelihood
				Wald statistic	Inference (5% signif.)	
1 Cobb-Douglas	unrestricted	0.718	0.116	5.91	reject	127.7
2 Cobb-Douglas	with CRTS	0.706	0.119	6.10	reject	126.3
3 Translog	unrestricted	0.708	0.109	7.18	reject	143.9
4 Translog #	with CRTS	0.751	0.107	5.44	reject	141.1
5 Translog	with local concavity	0.708	0.109	7.18	reject	143.8
6 Translog	CRTS & constant *wr*					
a. setting *wr* wage ratio = 1.0		0.753	0.106	5.43	reject	141.0
b. setting *wr* wage ratio = 0.6		0.741	0.110	5.59	reject	139.2

Note: In regression models 1 to 5, the conversion factor between female and equivalent male labour units is based on the actual value of the female to male wage ratio (*wr*) observed in a specific sample village. The average value of this ratio in this sample is 0.85 (Table 6.1b). In regression models 6(a) and 6(b) the observed *wr* value is replaced with a constant value (1 or 0.6, as indicated) for the entire sample of households.

\# Regression 4 is the specification corresponding to the full results of Table 6.9.

In particular it is heartening that the estimate of θ is quite insensitive to the conversion factor between male and equivalent female workdays. In the main estimation results reported in Sections 6.3 to 6.5 of this chapter, the conversion factor used is the ratio of the female to male wage rate (*wr*) in each sample village. The mean value of *wr* is 0.85 with a range from 0.64 to 1. In the results reported under Model number 6 of Table 6.11, the TL equation has been re-run by replacing the actual reported value of *wr* in each sample village with an exogenously specified value. Whether this exogenously specified common value of *wr* is 0.6 or 1, the estimate of θ changes only marginally. More importantly, the result that the estimate of θ is significantly less than 1 is not altered by changing the value of *wr*. Therefore, as long as a linear specification implying perfect substitutability between male and female labour is imposed *a priori*, the specific rate at which female workdays is converted to equivalent male workdays does not matter in the tests for heterogeneity between family and hired labour. It does not affect the main result that family and hired labour for both sexes are perfect substitutes with a constant higher productivity of family labour.

6.7 Elasticities of Substitution with a Linear Labour Composite

A complete set of the estimates of the AES, the HEC and the factor demand elasticities for the TL specification with the linear heterogeneous labour composite is presented in Table 6.12. These are computed at the geometric mean of Sample II using the regression parameter estimates of Model C2 in Table 6.9. Ignoring the own AES (values on the main diagonal of Panel (a)), all other AES are positive, with one exception. The negative AES between bullock power and material inputs indicates these two inputs are *p*-complements. An increase in the price of bullock power reduces the input demand for material inputs. Amongst the set of positive AES, the AES between bullock power and effective labour is close to one; but all others are much smaller than one. These values are quite different from previous estimates reported for Nepalese agriculture. For instance, in the set of AES reported for land, labour, bullock and fertilizers in Hamal (1991), all values are positive, and are either higher than or close to one.[45]

The matrix of factor demand elasticities derived from the estimated AES is given in Panel (b) of Table 6.12.[46] Labour is the most price elastic input with an own price elasticity of -0.463. This estimate seems to be on the high side, but is within the range of estimates observed for other countries – Bapna, *et. al.* (1984) and Singh Squire and Strauss (1986c1).

Table 6.12 Elasticities of Substitution and Input Demand
(with Le = F + θH effective labour composite)

(a) Allen partial elasticities of substitution

	Effec. labour	Effec. land	Bullocks	Materials
Effective labour	-1.415	0.663	1.076	0.533
Effective land		-0.493	0.173	0.092
Bullocks			-4.801	-0.662
Materials				-1.523

(b) factor demand elasticities

	Effec. labour	Effec. land	Bullocks	Materials
Effective labour	-0.463	0.322	0.081	0.059
Effective land	0.217	-0.239	0.013	0.010
Bullocks	0.352	0.084	-0.363	-0.074
Materials	0.174	0.045	-0.050	-0.169

(c) Hicksian elasticities of complementarity

	Effec. labour	Effec. Land	Bullocks	Materials
Effective labour	-2.541	1.921	-2.117	0.502
Effective land		-1.947	3.133	0.729
Bullocks			-6.549	-3.044
Materials				-2.605

Source: Computed from parameter estimates of Table 6.9 at the mean of the data.

The estimated HEC values reported in Panel (c) show that effective labour is a q-complement of land and material inputs. Increased usage of the latter two inputs increases the marginal product of labour. But human labour and animal labour are q-substitutes. Land is a q-complement for all other inputs, which is to be expected. In comparison to the HEC's computed for Sample I in Table 6.7, there is a reversal in sign in the HEC between effective human labour and material inputs. The positive HEC between human labour and material inputs in Table 6.12 is a more reasonable result than the negative HEC in Table 6.7. In both tables, however, the anomalous result of a negative HEC between bullock power and material inputs remains.

6.8 Summary

The production function estimation results in this chapter constitute the first step in the sequential estimation strategy of a farm household model that allows for labour heterogeneity. This first step estimation had two objectives: to test for heterogeneity between family and hired labour, and to generate a complete set of the factor demand elasticities and elasticities of substitution to describe the production technology.

The test for labour heterogeneity in itself had two components: a test of the separability of the labour inputs in the production function, and a test for the preferred specification of an aggregator function that converts family and hired labour into a composite labour variable, measured in effective units. The test for the separability of hired and family labour was based on a translog function specified with family and hired labour as distinct inputs, using a sample of households that used both family and hired labour inputs in crop production. The results showed that family and hired labour are only weakly separable from the other three main inputs – land, bullock power and material inputs. More restrictive types of labour separability are not accepted; and in particular the Cobb-Douglas form is strongly rejected.

The parameter estimates with family and hired labour as distinct inputs show these two labour categories are very close substitutes in both a price and quantity sense. At the sample mean, the estimated partial Allen elasticity of substitution between family and hired labour is positive and large (greater than 10). The Hicksian elasticity of complementarity is negative. The negative HEC implies that the marginal product of one type of labour is reduced through increased application of the other.

The labour separability results make it feasible to embed the test for labour heterogeneity in a production function specification that compares alternative ways of aggregating family and hired labour into a composite effective labour (Le) input. Five different effective labour functions, allowing for different values of the elasticity of substitution and the marginal rates of technical substitution between family and hired labour inputs, were estimated using a nested production function structure. The estimation was carried out over a larger sample of households that includes those cases not reporting any use of hired labour. The preferred effective labour aggregator function was the linear heterogeneous composite of the form $Le = F + \theta H$. The estimated value of θ was 0.751, with a standard error of 0.107. This estimate of θ is significantly less than one, implying that, although family and hired labour are perfect substitutes, they are not equivalent in efficiency units. There is a constant difference in their efficiency, with family labour being more productive.

The robustness of the result that θ is significantly less than one was checked through sensitivity analysis, allowing for alternative functional specifications and parametric restrictions, as well as by varying the rate at which the workdays of female labourers was converted into equivalent male workdays. These changes affect the estimate of θ only slightly and do not overturn the result that the estimate of θ is significantly less than one.

The finding of a higher productivity of family labour in own farm crop production in the *tarai* region on Nepal is consistent with the results of Foster (1994) and Deolalikar and Vijverberg (1983) with Indian data. Both of these previous studies, however, found evidence for imperfect substitution between family and hired labour, which is a result that differs from this study.

The complete set of the estimated production function parameters (Table 6.9) and the matrix of the various elasticities of input substitution derived from the estimated parameters (Table 6.12) are quite reasonable. All but one of the estimated AES are less than one indicating again that the simpler Cobb-Douglas specification is inappropriate for this data set. The estimated HEC values show effective labour is a q-complement for land and material inputs but a q-substitute for bullock power.

Finally, this chapter also indicates that there is considerable scope for functional mis-specification in treating different types of labour as distinct inputs in situations where the data contains many cases with a zero value for some particular labour categories. (This issue is taken up further in Appendix 6 below). A nested production function structure with an effective labour variable clearly appears to be superior to a specification that treats family and hired labour as distinct inputs.

Appendix 6 The Translog Specification with Family and Hired Labour as Independent Inputs

This appendix presents the translog production function estimation results with family and hired labour as distinct inputs, using Sample II which contains a large proportion of households that do not use any hired labour. A conventional approach to resolving the problem of zero values of some inputs in the translog specification has been to convert the zero values on all inputs to a value of one (or another small positive number).[47] An adjustment of this type seems particularly unsuitable with the data in Sample II because almost 60% of the households will have a zero level of hired labour input.

This presumption is clearly borne out by the estimation results given in Appendix Table 6A.1 for the translog and Cobb-Douglas specification with family and hired labour as distinct inputs. The other main inputs are as specified before – effective land, bullock days and material inputs.[48] The parametric estimates, especially for the labour inputs, in Appendix Table 6A.1 vary greatly from the specification that uses the Le = F + θH linear composite for effective labour, or even the homogenous labour composite. In the Cobb-Douglas estimates of Model 6 in Appendix Table 6A.1, the output elasticity with respect to family labour (α_1) is less than 0.06. The elasticity parameter with respect to hired labour is less by a factor of ten, and is not significantly different from zero. The effect of these parameter estimates is that the marginal product of family labour at the mean of the data is now estimated to be only Rs. 3.05 per male workday while that for hired labour is Rs. 0.59 per day. These are very unreasonable estimates (given that wage rates are about Rs. 9.9 per male workday), and they differ markedly from the marginal products computed with the homogeneous and linear composites in Table 6.10.

The translog specification (Model 5 in Appendix Table 6A.1) gives more reasonable results than the Cobb-Douglas, but anomalies remain. The elasticity with respect to hired labour (0.032) is considerably smaller than compared to family labour (0.258) at the mean of the Sample II data. At this mean point, the estimated marginal product of family labour (Rs. 13.86) is reasonable, but the marginal product of hired labour (Rs. 3.11) is considerably underestimated. These estimates of the marginal products and several parameters vary considerably from the corresponding estimates of the translog specification with family and hired labour as distinct inputs estimated over Sample I (Table 6.3). The latter results (which did not involve arbitrary conversion of zero values of inputs) are more reasonable.

Appendix Table 6A.1   Translog Production Function with Family and Hired Labour as Independent Inputs

Dependent Variable :	Log real composite farm output			
Data subset :	Sample II		(N = 679)	
Prior Restrictions :	Constant returns to scale			
Estimation Method :	Non-linear least squares			

		Model 5 # WS and local concavity		Model 6 Cobb-Douglas	
Additional restrictions:					
Variable		*coefficient*	*t- ratio*	*coefficient*	*t- ratio*
Family labour (F)	$\alpha 1$	0.258	8.14	0.0567	4.33
Hired labour (H)	$\alpha 2$	0.0328	4.75	0.0054	0.87
Effective land (Ae)	$\alpha 3$	0.501	10.37	0.753	24.90
Bullocks (B)	$\alpha 4$	0.093	4.44	0.055	3.32
Material inputs (M)	$\alpha 5$	0.115	4.47	0.129	5.52
Education of Head		0.006	1.12	0.0045	0.74
Big Farm dummy		0.003	0.11	-0.014	0.63
Second-order coefficients (x 10)					
FxF	$\alpha 11$	0.475	5.70		
HxH	$\alpha 22$	0.245	4.61		
FxH	$\alpha 12$	-2.110	3.58		
AexAe	$\alpha 33$	-1.290	1.57		
BxB	$\alpha 44$	0.360	2.41		
MxM	$\alpha 55$	0.715	1.70		
FxAe	$\alpha 13$	0.590	2.00		
HxAe	$\alpha 23$	0.070	1.89		
FxB	$\alpha 14$	-0.510	2.52		
HxB	$\alpha 24$	-0.060	2.36		
FxM	$\alpha 15$	-0.357	1.69		
HxM	$\alpha 25$	-0.045	1.63		
AexB	$\alpha 34$	0.569	1.68		
AexM	$\alpha 35$	0.047	0.98		
BxM	$\alpha 45$	-0.359	1.33		
R-Square (observed & predicted)		0.967		0.963	
Residual sum of squares		24.1		26.3	
Log Likelihood		133.2		105.6	
Wald test statistic for Cobb-Douglas form			60.1	(7 df)	*reject*
Estimated marginal product value at mean (Rs. per day)					
family labour		13.86		3.05	
hired labour		3.61		0.59	

WS = weak separability of the two labour inputs in the production function.

In the translog specification with scaled data, the production elasticity parameters (α's) are sensitive to the point of scaling of the sample data (Boisvert, 1982). If the hired labour variable is scaled to the mean, not of Sample II with the arbitrary conversion of zero values to one, but to the mean of Sample I (which has only those households with positive inputs of both family and hired labour), then the estimated marginal products become reasonable. At the mean of the Sample I data, the estimated marginal product is Rs. 14.46 for family labour and Rs. 10.97 for hired labour, using the estimated parameters of the translog specification of Model 5 in Appendix Table 6A.1. While the production technology of farms that employ both types of labour appear to be well represented by this specification, it however does not reflect well the common technology shared with the predominant group of farms not using any hired labour.

Note that despite the unreasonable parameter estimates with family and hired labour as distinct inputs, the goodness of fit criteria for Models 5 and 6 are very close to that for the specification with a composite effective labour function. In Model C2 in Table 6.9 the residual sum of squares is 23.71 and in the translog specification of Appendix Table 6A.1, the residual sum of squares increases only marginally to 24.1. There is no change in the R-square between observed and predicted values. It appears the substantial model mis-specification with family and hired labour treated as distinct inputs, with zero values converted to one's, could be masked by the standard goodness of fit criteria, especially in a non-linear estimation procedure.[49]

In summary, the estimation results in Appendix Table 6A.1 clearly illustrate the problem that can occur when the *ad hoc* procedure for converting zero values on inputs in situations when such zero values occur in a large proportion of the sample. The results with the effective labour aggregator function clearly are more economically meaningful even though statistically the model fit may not be very different.

Notes

1   See also the spirited defence of the primal approach to production function estimation offered in Mundlak (1996). He argues that estimating dual specifications with prices assumed exogenous at the firm level does not utilize all the available information, and the loss in efficiency can be sizeable.

2   The dependant variable in the aggregate production function is a real index of composite crop output created by deflating the nominal value of output by a village-specific producer price index. The subtotals for family and hired labour add up the

male and female labour days, after converting female workdays into equivalent male days, using the observed female to male wage ratio. For instance, total hired labour days = hired male days + wr (times) hired female days, where wr is the ratio of the wage rate of the female labourer to the male labourer's wage rate that is observed in a particular sample village. The total family labour input is computed similarly, using the wr observed for hired labour wage rates. See Section 5.3.1 in Chapter 5 for further discussion and some justification of this simple method for aggregating male and female labour inputs.

3   The first of these adjustments in effect re-classifies, in terms of the operating land strata definitions adopted in the sample design of the MPBHS, marginal farm households with an operated land size of less than 0.1 hectares as landless households. The MPHBS strata limits are arbitrary since there is no clear economic distinction between a completely landless household and an almost landless one. It is appropriate to exclude from the production function estimation those households with very small plots of land that could be just extensions of the homestead plot. Similarly, farm households that do not report any family labour input in crop production are also unusual. Since some of the composite labour functions used in Section 6.4 of this chapter rely on a ratio format of the two types of labour, and since most households use some family labour, the cases without any family labour input are dropped from the sample. This ensures that the ratio of hired to family labour is defined for all included cases. The same sample is used for all other functional specification in order to have a consistent comparison of the results. The adjustments to the sample size due to these two arbitrary cut-offs are minor. Only 34 additional households (4.8%) are excluded from Sample II for these reasons.

4   Given the specific translog functional form chosen for the production function estimated in Section 6.3, it is not just desirable but also necessary to restrict the data set to households with non-zero values for family and hired labour inputs since the regressors are defined in terms of logarithm values.

5   These percentages are the proportion of households in the big farm category reported in Tables 6.1a and 6.1b. The MPHBS sample design, as described in Appendix Table 5A.1, had four operated land holding strata. The proportion of farm households classified in the big farm category for this study is based on re-classifying the large and medium farm size strata of the MPBHS sample design as big farms. Similarly, the smaller other two strata have also been collapsed into a single small farm category for the estimation work of this chapter and Chapter 7. Consequently, the dividing line between small and big farms in Tables 6.1a and 6.1b (and in all the estimation results reported in this chapter) is 2.73 hectares.

6   The share of female labour in total labour input (from both family and hired sources) is almost the same in Sample I and Sample II - about one third. The respective proportions in the family and hired categories are different. In Sample I about 28% of total family labour is female workdays but the corresponding proportion is 39% for hired labour. In Sample II the differences in gender composition are smaller - about 34% of family labour is female and about 39% of hired labour is female.

7   The four land operating strata adopted in the original MPHBS sample design have been collapsed into two size categories only. (See Endnote 5 above). So a single binary variable captures the farm size dummy variable, which takes the value of one for a farm household operating more than 2.73 hectares of land (big farm). Similarly, the 32 sample village clusters in the five districts are represented by dummy variables to capture underlying land quality differences through fixed effect intercepts.

8   These parametric restrictions are based on the treatment of the translog flexible form as a second order approximation to an unknown arbitrary production function. Denny and Fuss (1977) call this an "approximate test" for separability. When the translog is interpreted as an exact production relationship the separability tests involve additional restrictions on the parameters, as in Berndt and Christensen (1973b). Denny and Fuss (1977: 405) discuss the differences between the approximate and exact tests for separability and recommend that the former be used as a test for separability because the latter involves a joint test for separability *and* a specific functional form, and hence is unduly restrictive.

9   The restrictions of Equation 6.6a can be re-written as

$$\alpha_1\alpha_{23} = \alpha_2\alpha_{13} \; ; \; \alpha_1\alpha_{24} = \alpha_2\alpha_{14} \; ; \; \alpha_1\alpha_{25} = \alpha_2\alpha_{15}$$

which are always satisfied when Equation 6.6b holds. Hence partial strong separability is a sufficient but not necessary condition for weak approximate separability.

10   Another term for complete strong separability is factor wise separability (Chambers, 1988: 46-47). The weak and partial strong separability conditions are restrictions about the equality of the Allen elasticity of substitution between inputs in different partitions of the input list (Berndt and Christensen, 1973a). The Cobb-Douglas form not only imposes equality of the AES between all input pairs but also restricts the value of this common AES to 1.

11   The Wald Chi-square test statistic is used for the tests on separability, even though the equations are estimated by OLS, because the tests in some instances involve non-linear restrictions on the parameters.

12   The CRTS restrictions in Equation 6.5 are imposed on the five X variables only. The parametric restriction implied by CRTS in the translog equation are:

$$\sum_{i=1}^{5} \alpha_i = 1; \text{ and } \sum_{j=1}^{5} \alpha_{ij} = 0 \quad \text{for each } i = 1 \text{ to } 5.$$

13   If the *p*-value is larger than the significance level used for the test of the null hypothesis, the null is *not* rejected.

14   Furthermore, the CRTS restricted estimates have a smaller variance-covariance matrix than the unrestricted estimates; so if the restrictions are accepted, it is desirable to use the restricted specification (Greene, 1997: 342).

15   The non-rejection of the null of weak separability of the two types of labour in Test B of Table 6.2 is not conditional on the prior restriction of constant returns to scale. In Test A, WS is accepted even without imposing CRTS. In Test B if WS is imposed at the first level and CRTS tested in the second level, the same overall test result occurs. The only difference is in the levels of significance of the tests at individual levels.

16   The translog regression results reported in Table 6.3, as well as in all other subsequent tables in this chapter are incomplete results. The coefficients for other variables included in these regressions are not shown. The main set of such excluded variables was the 31 intercept dummy variables for the sample village clusters (village *panchayats*). The combined set of these dummy variables in the regressions results are very significant, indicating substantial unobserved regional heterogeneity in crop production at the level of the sample village clusters.

17   Concavity of the translog function at specific positions can be tested by verifying the negative-semi-definiteness of the Hessian matrix of second derivatives (Berndt and Christensen, 1973b).

18   Local concavity imposes concavity at a point of approximation of the TL function - such as the geometric mean of the sample data. Global concavity requires the function to be concave at every point and this often reduces the flexibility of the translog specification (Diewert and Wales, 1987). The procedure for estimating the TL equation with global concavity restrictions follows the Cholesky decomposition approach of Jorgenson and Fraumeni (1981). The procedure adopted for imposing local concavity restrictions at the geometric mean of the data follows Ryan and Wales (1998).

19   The data underlying the parameter estimates in Table 6.3 have been scaled at the geometric mean. Therefore the α_i represent the input elasticity at the geometric mean and are estimates of the share of input i in the value of total output under competitive market conditions (Boisvert, 1982).

20   Hicks (1970: 294).

21   The Hicksian elasticity of complementarity (HEC) was introduced in Hicks (1970). It is the inverse of the Hicksian elasticity of substitution (HES) defined in the *Theory of Wages* (Hicks, 1964). Sato and Koizumi (1973) have shown it is more appropriate to work with the HEC in a multi-factor setting since the HEC and AES represent the duality between the production and cost functions. In a two-factor setting with constant returns to scale, the AES and HES are identical (Hicks, 1970), or equivalently the AES is the inverse of the HEC (Sato and Koizumi, 1973.)

22   These elasticities are computed by applying Equations 6.8 (a,b,c) to derive their specific form in terms of the estimated parameters of a translog production function. These derivations are fairly involved and hence are not repeated here. The computation of the AES for a translog function is detailed in Berndt and Christensen (1973b); and the derivation of the HEC and DES for the translog is given in Squires and Tabor (1994).

23   Note that the calculation of the AES from production function parameters involves inverting the bordered Hessian matrix of second order derivatives. (See Eq. 6.8a). Hence, if any of the production function parameters have large standard errors, the resulting estimates of the AES are imprecise, with standard errors that are not readily computable (Squires and Tabor, 1994).

24   The Wald Chi-square statistic for the test that the ratio of marginal products is equal to one is 15.3, with 1 degree of freedom. The *p*-value is 0.0001.

25   The sample subset that uses only family labour consists of those farm households that are in Sample I but not in Sample II. The number of farm households in this subset is 400 (the 679 in Sample II minus the 279 in Sample I).

26   It is awkward to carry out the tests for common production function parameters based on the nested production function structure with an effective labour nest, as specified in Section 6.4 to follow. This occurs not only because of the non-linear estimation procedures for which the F test is only approximate (Greene, 1997: 360); but also because the parameter set between samples using hired and family labour and those using only family labour will differ in the nested production function.

27   Adopting such a procedure leads to substantial downward bias in the linear coefficients of hired labour (α_2 in Table 6.3) and, hence in the implied ratio of marginal productivity of family and hired labour. See Appendix 6 for a fuller discussion, and in particular the results in Appendix Table 6A.1.

28   The traditional form of the CES function is $\gamma\,[\delta{*}F^{-\rho} + (1-\delta){*}H^{-\rho}]^{-1/\rho}$, where γ is the efficiency parameter and δ the distribution parameter (Arrow, *et al.* 1961). This can be re-specified as $\gamma`\,[F^{-\rho} + \theta{*}H^{-\rho}]^{-1/\rho}$ where $\gamma` = (\delta^{-1/\rho})\,\gamma$ and $\theta = (1-\delta)/\delta$. Since $\gamma`$ is

just a scaling factor which gets converted to a constant when taking logs, the CES form can be written as in C3. Similarly in the GL specification of C4, the coefficient for F can also be normalized to one.

29   A multi-factor standalone (non-nested) CES production function has the undesirable property that the elasticity of substitution for every pair of inputs is exactly the same (Uzawa, 1962). The second order conditions for the concavity of a multi-factor production function with n inputs requires that at least $(n-1)$ of the Allen partial elasticities of substitution must be positive (Sato and Koizumi, 1973). Therefore in a multi-factor CES, all inputs must be p-substitutes with positive AES.

30   From Equation 6.8b, the sign of HEC_{ij} depends only on the cross derivative f_{ij}. In the CES nest of C.3, the sign of f_{FH} is determined by the sign of $(1 + \rho)$, which is always positive when $\rho \geq -1$. Moreover, in the two input CES case considered here, the HEC between F and H is given directly by $(1 + \rho)$ since the HEC is the inverse of AES in a production function with only two inputs (Sato and Koizumi, 1973).

31   In a nested production function structure it is necessary to distinguish the AES between F and H in the $g(.)$ function for the composite labour input (Le), and the AES between F and H in the main or primary level production function. This formula computes the AES_{FH} in the composite labour function of the Ratio format (Equation 6.9. C5).

32   The restriction $|\mu| \leq 1$ is needed to maintain concavity of the ratio aggregator function.

33   Equation 6.10 gives the formula for the AES between F and H in the Le function of the GL specification (Deolalikar and Vijverberg, 1987). See Endnote 31 above.

34   Since the dependent variable in the production function regressions is aggregate output, allowing for $\kappa < 1$ is also an implicit control for variations in the cropping pattern that may arise from differences in the level of multiple cropping, in contrast to differences in the physical land area across farms.

35   This is a reasonable result because upland does not generally have a high water retention rate; hence unirrigated upland dependent on rainfall may be just as productive as upland which has a more dependable irrigation source.

36   In logarithmic form for the overall translog function, the specification for the effective land variable becomes $\log (Ae) = \log (A_{c \cdot Pi} + \gamma_1 A_{c \cdot Pu} + \gamma_2 A_{c \cdot U}) + \kappa \log (CRINT)$.

37   In the restricted version of the CES nest, the boundary limit on $\rho \geq -1$ is reached, so this is equivalent to estimating the CES nest with $\rho = -1$ imposed *a priori*.

38   Based on the log likelihood values reported for Models C.1 and C.2 in Table 6.8, the Likelihood Ratio test statistic is equal to $-2 x (138.56 - 141.06) = 5.02$. The critical value of the Chi-square test statistic with 1 degree of freedom is 3.8. Therefore this test also shows that the sample data is not consistent with the restriction that $\theta = 1$.

39   A positive μ implies that for a given level of aggregate labour $(F + H)$, a higher ratio of family labour in this total increases the level of effective labour. A positive μ also means the AES_{FH} and HEC_{FH} are positive, making F and H p-substitutes and q-complements.

40   The ratio specification and the linear composite have the same number of parameters. Therefore a smaller residual variance, which leads to a higher value of the log likelihood, is equivalent to that model being preferred on several model selection criteria, such as the Akaike Information Criteria, which depend only on the residual sum of squares and degrees of freedom (White, 1993: 16).

41   For the unrestricted CES, the estimated of ρ is less than minus one, but not significantly different from -1. This reduces the CES to a linear composite with an infinite AES.

42   Frisvold's estimate of μ was 0.24. At the sample mean this leads to an AES between F and H of 1.71 in the effective labour function, and 0.33 in the main production function (Frisvold, 1994: 230-231).

43   These estimated relative differences in unit productivity of unirrigated paddy land and upland *vis-a-vis* irrigated paddy land appear to be smaller than expected, given the primacy of paddy cultivation in the *tarai* region of Nepal. One factor contributing to the low relative productivity differences is that the land input variable is already measured in terms of gross harvested area and not in terms of physical units of land. If irrigated land tends to be multi-cropped more often, and if the yield in the second or subsequent crop grown outside of the normal seasonal rotation is likely to low, this will depress the average productivity of irrigated land (compared to the yield of unirrigated land which is cultivated only during the normal season). Secondly, the MPHBS data does not contain details on the quality and reliability of irrigation facilities on the sample of farms. Since the timing of irrigated water supply is the critical issue in boosting crop yields, it may be that differences in the yield between irrigated and unirrigated land may be more pronounced in years where there is general drought due, say, to the monsoon failure. In a normal year, as in our sample, when the monsoon rainfall was adequate, the relative differences in yields could be less extreme if adequate rainfall has occurred at critical times in the growing season.

44   As described in Chapter 5, the wage rates are compiled from the actual wage payments made by households that report hiring in labour for each sample village cluster. Since hiring in farm labour is likely to be more common in the peak labour demand seasons, these wage rates will tend to be higher than an annual average of the prevailing market wage rates for hired labour, when there is seasonal fluctuation in the wage rates.

45   Hamal's estimates are not specific to the *tarai* region agriculture, but for all of Nepalese agriculture, using a time series estimation technique with aggregate data. Using a translog cost function specified with four main inputs (land, homogenous labour, bullock power and chemical fertilizers) he computes the Allen partial elasticity of substitution between these inputs for various time sub-periods. All of the estimated AES are positive – implying all pairs of inputs are *q*-substitutes (Hamal, 1991: 142).

46   The factor demand elasticities (ε_{ij}) are readily derived from the AES since $\varepsilon_{ij} = \sigma_{ij} \cdot s_j$, where s_j is the share of input j in the total value of production (Sato and Koizumi, 1973: 48). Since the translog regression parameters are estimated with data scaled to the geometric mean, the value of the s_j's at the mean are just the estimates of the input elasticity parameter (α_j's)

47   Jacoby (1993) is one example of this commonly used *ad hoc* procedure.

48   The prior restriction of constant returns to scale is imposed in both models. In the translog specification (Model 5 in Appendix Table 6A.1) the data are scaled to the geometric mean which results from the conversion of the zero values of hired labour to one. Weak separability and local concavity are also imposed.

49   If OLS estimation could be used then the wider variety of mis-specification diagnostics tests (such as the RESET) could adequately differentiate between models that have family and hired labour as distinct inputs (such as Models 5 and 6 in Appendix Table 6A.1) and models with an effective labour aggregator function. Many standard specification tests are not applicable for non-linear estimation (Greene, 1997: 459-60); and so have not been applied here to discriminate in a more formal way between the specification of Model 5 in Appendix Table 6.A1 and Model C2 in Table 6.9.

7 Labour Supply Estimation

7.1 Introduction and Motivation

The results in Chapter 6, of a production function based test for the heterogeneity of family and hired labour as farm inputs in the *tarai* region of Nepal, indicated that family and hired labour are perfect substitutes in crop production but with differing productivity. This finding of a linear composite form of labour heterogeneity has important implications for the appropriate methodology as well as model specification in estimating the labour supply component of the farm household model. As discussed in Chapter 3, the methodological implication of a linear form of labour heterogeneity is that the farm household model is still *recursive* in its production and consumption decisions. The constant efficiency difference between hired and family labour, denoted by θ, leads to a model structure wherein the effective wage rates faced by household members will differ according to the labour market exposure of the household in the hired labour market – i.e. whether the household is a net buyer or net seller of labour. The efficiency difference, however, is independent of the levels of the labour inputs, and of the other inputs used on the farm (because the production function was shown to be separable in the labour inputs). Hence, the effective wage rate faced by family labour when applied to the farm is still parametrically given to the farm household. Consequently, the labour supply estimation can be carried out separately from the production side of the model, but with the necessary adjustments to the observed market wage rates for the difference in productivity represented by θ.

This chapter presents the estimation procedure and results for the labour supply behaviour of farm household members taking into account the observed heterogeneity between family and hired labour in farm production. The resulting set of estimates of the parameters of the labour supply function completes the description of the behavioural response of the farm household. Another equally important motivation for this chapter is to provide independent corroboration of the result derived in Chapter 6 that family and hired labour are heterogeneous inputs in crop cultivation.

145

The production function based test for $\theta < 1$ reported in Chapter 6 is robust with respect to several alternative functional specifications and parametric restrictions (Table 6.11). While this is strong evidence for labour heterogeneity, it is not conclusive in the sense that it does not rule out alternative explanations of why the estimated θ could be less than one in the production function regression results. The main concerns are of data aggregation bias and unobserved quality differences in family and hired labour inputs that are independent of their family or hired status.

The farm management data utilized in the regression analyses of Chapter 6 is given only at the aggregate household level in the MPBHS. Individual characteristics of persons who supply the family and hired labour input on a particular farm are not observed. There could be some unobservable quality differences between family and hired labour that give rise to a difference in their marginal productivity independently of the distinction between family and hired labour categories. For instance, in a particular household family labour may consist solely of prime age workers, while the hired labour input may be of young teenagers or older adults. Similarly, education levels of family and hired labour may be different. Another source of unknown bias in the estimate of θ may arise from the aggregated nature of the farm level production function estimated in Chapter 6, since production technology and labour productivity can differ across crops. Therefore it is important to find *additional evidence* for the higher marginal productivity of family labour independently of the production function estimation. The labour supply regressions presented in this Chapter provide a mechanism for such an independent verification.

It is feasible to test whether the observed labour supply behaviour is consistent with the heterogeneity between family and hired labour detected in the production function estimation by comparing alternative labour supply model specifications. A labour supply model that equates the opportunity cost of family labour in all households to the observed market wage rate for off farm work is consistent with family and hired labour being homogeneous production inputs. Model specifications using effective wage rates (based on θ) that vary according to the labour market exposure of the household are consistent with labour heterogeneity. Standard diagnostics for model selection can be used to check whether the common wage labour supply model performs better than the varying effective wage model. If these model selection tests find in favour of the varying wage model, this result can be interpreted as conforming to the production function based result on efficiency differences between family and hired labour as inputs in farm cultivation.[1]

The superior performance of a labour supply model with varying effective wage rates does not necessarily prove there must be an efficiency difference between family and hired labour in the production function. Such a result could be consistent with other explanations that lead to differences in the effective wage rate for family labour applicable to own farm work and the wage rate applicable for work on the hired labour market. A wage gap of this type can also arise from differences in preferences over working on one's own farm and working off farm (Lopez, 1984), or from fixed costs to seeking off farm work (Cogan, 1981). The tests of the labour supply model specifications in this Chapter are not designed to discriminate between alternative sources or explanations for varying effective wages. They only discriminate between models relying on a common market wage and models that specify varying effective wage rates, based on whether households are net buyers or sellers of labour.

If the labour supply behaviour conforms to the production function based evidence for labour heterogeneity, this provides stronger evidence for a genuine efficiency difference between family and hired labour. Were the results in Chapter 6 due solely to other unobserved missing variables in the production function estimations, it would be an unlikely coincidence if these missing variables would also lead to a labour supply model characterized by higher effective wage rates for family members.[2]

A secondary objective of this chapter is to obtain accurate estimates of the parameters of interest in the labour supply functions (i.e. wage and income elasticities) by correctly specifying the appropriate effective wage that defines the leisure/consumption equilibrium of different households. It is of some interest to check how these elasticities will differ between specifications that recognize labour heterogeneity (and hence the wage gap) and those that treat both types of labour as homogeneous inputs.

In the following sections, the next (7.2) briefly discusses how the household's optimal labour supply/leisure demand conditions are affected by labour heterogeneity; and, in particular, how the effective wage rates at the equilibrium are related to the θ parameter. Section 7.3 provides a short summary of the individual and household level data used in the labour supply regression estimations. Section 7.4 discusses the specification of the alternative models and the related model identification issues. Section 7.5 provides the labour supply regression results for male household members, including model selection tests and estimates of the wage and income elasticities of labour supply from alternative models. Section 7.6 gives a similar set of results for the labour supply of female household members. Section 7.7 provides a summary of the main results of this chapter.

7.2 Labour Supply Implications of Linear Heterogeneity

The effect of a linear composite function for effective labour ($Le = F + \theta H$ with $\theta < 1$) is that one unit of family labour when applied to the family farm can substitute for $1/\theta$ units of hired labour, without affecting output. If w denotes the market wage rate paid per unit of hired labour, for the farm household which hires in some labour the effective wage rate that can be applied to family labour devoted to the family farm is w/θ. (One unit of F substitutes for $1/\theta$ units of H for which the hired wage cost is w/θ). Moreover, since the $1/\theta$ conversion factor is independent of the actual levels of family and hired labour (and other inputs) applied on the farm, there is a constant difference in the marginal product of hired labour *vis-à-vis* family labour. The marginal rate of technical substitution between family and hired labour (which is given by the ratio of their marginal products) is constant for all households. The first order conditions for the optimal levels of family and hired labour inputs can be derived as if the household faced a wage rate of w for hired labour and an internal wage rate of w/θ for family labour when applied to its own farm (see Section 3.5 in Chapter 3). Even though w/θ is a shadow wage rate that applies only to an artificial internal labour market for family labour within the household, the family labour supply behaviour can be modelled as if the household were to take the w/θ wage as parametrically given (assuming that the market wage rate w itself is exogenous).

This maintains the recursive nature of the farm household model whereby labour supply decisions can be modelled independently of other production input choices. Consequently, the traditional estimation strategy can be followed, where the production function and labour supply/ consumer demand systems are estimated separately, with a minor adjustment to market wage rates to derive the effective wage rates that represent the consumer-household's consumption/leisure equilibrium.

A second major implication of the linear effective labour aggregator function is that the efficiency difference between family and hired labour as production inputs affects the effective wage rate, and hence labour supply decisions, only of households that hire in labour (or, are at the margin of deciding to employ the first unit of hired labour). For a small farm household that supplies family labour to its own farm and also to the off farm hired labour market at a wage rate of w, the optimal family labour input on own farm cultivation is conditioned by the market wage rate, w, and is not affected by the θ parameter. Hence, the effect on the labour supply behaviour of a farm household of the efficiency difference between family and hired labour as production inputs depends on the net labour market position of that particular household.

The linear efficiency difference leads to three mutually exclusive household categories based on the net labour market position. If M represents the amount of family labour supplied on the hired labour market (at wage *w*) and H the amount of labour hired in (again at wage *w*), the three mutually exclusive household categories are:[3]

Category 1:   family labour hired out (M > 0), no labour hired in (H = 0). This category represents landless and small farmers who are net sellers of labour.

Category 2:   family labour not hired out (M = 0), and extra labour hired in (H > 0). This category represents big farmers who work on their own farms and also hire in labour.

Category 3:   neither family labour hired out (M = 0), nor extra labour hired in (H = 0). This category represents autarchic households who equate labour demand and supply on the family farm solely from family sources.

The possibility of both M > 0 and H > 0 is explicitly ruled out if $\theta < 1$ and the wage rates for hiring in and hiring out are the same. Since one unit of family labour is equal to $1/\theta$ units of hired labour, if the two wage rates are the same, the household always gains by transferring its labour from market wage work (activity M) to own farm work (activity F) since it can reduce hired labour demand by $1/\theta$ units for every unit of family labour so transferred.[4]

The prediction that, in the presence of a linear form of labour heterogeneity, with family labour being more productive than hired labour, farm households should not simultaneously hire in and hire out labour is a strong result. It can be readily tested with the sample data of this study. Table 7.1 confirms that indeed the sample households of the Nepal *tarai* region do not report simultaneous hiring in and hiring out of labour. Out of 686 farm operator households for which a matched set of labour demand and labour supply data could be assembled (from the subset of the MPBHS sample used in this study), only 22 households – about 3% – report using hired labour as well as some member of that household working on the off farm labour market. Most instances of simultaneously hiring in and hiring out are found among the smaller farm size categories, where the amounts of hired labour used are very small. Out of 199 large and medium sized farm households only three report simultaneous hiring in and hiring out.

Given that the MPHBS data refer to the entire annual cropping cycle, and given the very time specific nature of agricultural operations and the strict gender-related division of labour, this is a striking result, which is consistent with the type of labour heterogeneity detected in the production function estimation.

Table 7.1 Market Labour Supply Exposure of Sample Households

Operated land strata *	Sample size (N)	Total N with matched data	Number of households reporting:		
			Hiring in labour	Hiring out labour	Hiring in and out
(column 1)	(2)	(3)	(4)	(5)	(6)
Large	103	77	69	2	1
Medium	123	122	87	10	2
Small	227	214	89	62	8
Marginal	281	273	48	179	11
All cultivators	734	686	293	253	22
Landless	273	240	0	240	0
Total	1,007	926	293	493	22

Note: The total N reported in column 3 is the number of households for which both the labour supply and labour demand data could be matched from different parts of the MPHBS data records.
* These farm size strata were used in the stratified sample design of the MPHBS. See notes to Appendix Table5A.1 for the definition of the strata size limits.

The equilibrium conditions for labour demand and supply for each of the three analytically distinct household categories were illustrated in Figure 3.4 in Chapter 3 for the case where θ* was less than one at the optimum allocation of family and hired labour. The theoretical derivation in Chapter 3 was based on a general form of labour heterogeneity in a production function with a separable labour nest. The specific linear composite form of the labour nest indicated by the empirical results of Chapter 6 further simplifies the theoretical structure and estimation strategy. It is not necessary to compute θ^* at the optimum labour allocation for each sample household to define the effective wage rate applicable for the labour supply equilibrium. The applicable effective wage rate can be determined solely from the constant θ parameter estimated in Chapter 6 and the observed market wage rates.[5]

Figure 3.4 in Chapter 3 showed that the wage rates that define the equilibrium labour supply position on the labour supply curve are different for the three household types. For Category 1 (mainly landless and small farm) households, where total labour supply is greater than own farm demand ($M > 0$), both the implicit valuation of family labour and the marginal return to own labour in farm production are equal to the wage rate (w) received on the hired labour market (Fig. 3.4, Panel a). This result is independent of the value of θ. At the optimum labour allocation, the real cost of labour in all household activities is equalized to the market wage.

For Category 2 households (big farms where labour is hired in) the optimal use of hired labour equates the marginal product of hired labour to the hiring in wage rate (w); whereas the allocation of family labour to own farm production equates the marginal returns to w/θ (Fig. 3.4, Panel c).

For Category 3 (autarchic) households which neither hire in nor hire out any labour the equilibrium conditions set the effective wage rate for the labour supply equal to the marginal product of labour on the family farm. This common value for the real cost of labour and the marginal product, ω, lies between w and w/θ (Fig. 3.4, Panel b). ω is a household level shadow wage rate that can vary among the set of autarchic households even if they face the same market wage rate. For autarchic households, a small change in the market wage rate will have no effect on equilibrium labour supply or demand as long as the autarchic nature of the household is preserved – i.e. if the wage change is not large enough to make the autarchic household change its status into a Category 1 or Category 2 household. The underlying farm household model is not recursive for autarchic households even with a linear form of labour heterogeneity.

The effect of the lower efficiency of hired labour on own farm labour allocation decisions of the three different categories of farm households can be summarized in terms of the wage at which one unit of effective labour is available to each household type. This is illustrated in Figure 7.1, where labour is measured in effective units on the horizontal axis. As in Figure 3.4, the VV curve denotes the marginal rate of substitution between leisure and consumption for a representative individual worker in a farm household. The stepped bold line wSS'S" represents the supply price at which an effective unit of labour is available to the three different farm households. The classification of the labour market exposure of a farm household is based on the amount of labour required for own farm cultivation which determines where the demand for effective labour (or the schedule for the marginal product of effective labour) intersects with the labour supply curve of wSS'S''.

As drawn, Y^S reflects the demand for effective labour on a small farm that has extra labour to sell on the off farm labour market at wage w. For such a household, the price at which a unit of effective labour is available for own farm cultivation is also w (since units of family and effective labour are equivalent). When the farm size is large enough to require hired labour for cultivation, the supply price for a unit of effective labour becomes w/θ. Such a farm is represented by the Y^B schedule for the demand for effective labour. The large farm faces an effective wage rate of w/θ because a standard unit of hired labour at wage w represents only θ units of effective labour. An autarchic household that neither sells nor buys any labour is represented by the intersection of the marginal product of labour curve denoted as Y^A with the upward sloping section SS' of the supply curve. The marginal product of family labour, and hence its shadow price, for the autarchic household is between w and w/θ. Figure 7.1 indicates that if preferences for leisure and production technology are the same for all households, the sole basis for the classification of households according to their exposure in the hired labour market will be farm size, which determines the horizontal distance at which the marginal product curve intersects the supply curve. (As drawn, Ld^S, Ld^A, Ld^B represent the demand for effective labour in efficiency units on the representative small, autarchic and big farm households, respectively).

Figure 7.1 also shows that the effect of the θ factor in the productivity difference between family and hired labour is equivalent to a wage gap in the price at which an effective unit of labour is available to different farm size categories.[6]

The per hectare labour input in effective units applied to these three different farm categories will vary because the wage rate that each category faces is different. There is a rising supply price of labour only among the autarchic households. For the small and big farm households, although there is a gap in the effective wage they face, the supply of effective labour is available at a fixed price. Within the set of big farm households that rely on hired labour, there is no further difference in the wage cost: all big farms face the common wage of w/θ for one unit of effective labour.[7]

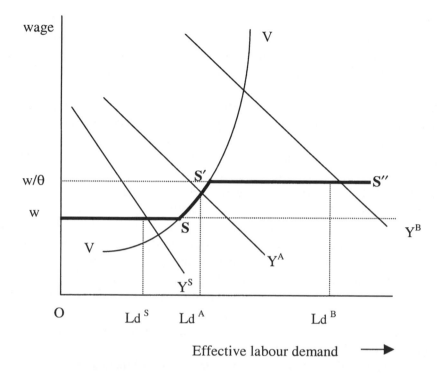

Figure 7.1 Supply Price of Agricultural Labour with Linear Heterogeneity

7.3 Data Summary

The main points related to the design and structure of the MPBHS data and variable definitions used in this study were discussed in Chapter 5. The specific data issues relevant to this chapter were the definition of labour supply adopted in the MPBHS coding manual, and the derivation of household specific market wages rates for male and female hired labour from the farm management module of the survey. (See Section 5.3 in Chapter 5 for details). This section provides some additional information about how the actual sample of individuals used for estimating the labour supply regressions, separately for male workers and female workers, is constructed. It also provides a summary description of the main regression variables of interest in both samples.

The labour supply regressions are based on the individual unit records in the MPBHS for economically active household members. This data records the monthly days of work in several different work categories for each of two six-monthly cropping season survey cycles. The full MPHBS data contains employment details for all economically active persons aged 10 or above. The sample used for the labour supply regressions in this chapter is further restricted in three ways to individuals who are: (i) aged 15 to 60, (ii) are in a familial relationship with the household head, and (iii) whose usual occupation codes are related to farm cultivation.

The age group restriction is imposed to ignore the occasional labour supply records for children aged 10 to 15 and the elderly. Also excluded are individuals not related to the household head. One important category thus excluded is the group of domestic and farm servants reported to reside in the household of their employer. The labour supply for such individuals would differ from that of family members in the household. The contractual arrangements for permanent or semi-permanent farm labourers can be complex, and their labour supply behaviour need not be related to an observed market wage rate for hired casual labourers.[8] Finally, the occupational code restrictions are imposed because the effective wage rates for family labour are to be created from the reported wage rates for hired labour used in farm cultivation. This wage rate for farm labour need not be a good proxy for the marginal returns to an extra unit of work for individuals in other village occupations – shopkeepers and caste based professions such as shoemakers, blacksmiths, etc. – whose labour supply details were also recorded in the MPHBS data.[9]

The final adjusted sample size available for the labour supply regressions consists of 2,542 person-season records for male family members and 2,288 person-season records for female family members. Given that both of the summer and winter season records are available for most individuals, this represents a sample size of about 2,500 economically active family members (male and female) from the 1,007 sample households of the selected five *tarai* region districts of the MPHBS.

In the case of female family members, extra data was also compiled for those individuals aged 15 to 60 who are reported as economically inactive. In order to counter potential sample selection bias in the regression results it is necessary to generate a sample data set that includes active and inactive family members. A total of 481 women aged 15 to 60 were reported to be economically inactive. When compared to the final sample of about 1,200 economically active women aged 15 to 60, this represents a non-participation rate of about 29%. There is no corresponding problem of potential sample selection bias in the sample of male family members. Only 26 cases of economically inactive men aged 15 to 60 were recorded.[10]

A summary description of the main variables used in the labour supply regressions for these three subsets of the data (male workers, female workers and female inactive family members) is given in Table 7.2. It also provides a breakdown of the data for the main variables by the three categories of households based on their labour market exposure. Table 7.2 shows that the average number of days worked is strongly related to labour market exposure. In both the male and female workers samples, average workdays is highest for Category 1 households (labour hire out) followed by Category 2 (labour hire in), and lowest in the autarchic (Category 3) households. The association between labour market exposure status and average farm size is also evident. For instance in the sample of male workers, individuals in the labour hiring out category come from landless and small farm households, with an average farm size of 0.88 hectares. Individuals in the autarchic category have an intermediate average farm size of 2.2 hectares. Average farm size in the labour hiring in category is 4.67 hectares. There is a similar relationship between per capita non-labour income and the labour market exposure categories, since this income category includes the imputed returns from land ownership (farm profit). Table 7.2 reveals a large variation in the non-labour income variable. This helps to accurately estimate the income effect on labour supply. The sample variation on real wage rates is also high, reflecting regional patterns. For instance, in the sample of female workers their real wage rate ranges from 2.63 to 7.43 (kilograms of paddy a day).[11]

7.4 Model Specification and Identification

In the second step of the farm household model estimation strategy, the specification of the labour supply equation, based on the estimate of θ obtained in the first step, is given by (from Equation 4.11 in Chapter 4):

(7.1) $Ls = \ell(w^*(\theta), \pi(\theta) + E, S, \Phi) + e$

where $w^*(\theta)$ $= w$ for Category 1 households
                  $= w/\theta$ for Category 2 households
                  $= \omega$ for Category 3 households
                        where $w \le \omega \le w/\theta$

$\pi(\theta) + E$ $=$ non-labour income, including farm profit
Φ $=$ the labour supply parameter set
e $=$ random error term.

Table 7.2 Labour Supply Data Summary

Variable	mean	std. dev.	min.	max.
Male Workers Sample (N = 2,542)				
Total labour supply in days [a] (all categories)	88.6	41.7	0	177
Category 1 Households (hire out) (N1 = 1,214)	108.5	38.7	0	177
Category 2 Households (hire in) (N2 = 813)	78.1	36.3	2.2	173
Category 3 Households (autarchic) (N3 = 515)	60.4	33.9	5	171
Per capita real non-labour income [b]	619	629.6	7	5,905
Category 1 Households (hire out labour)	255	242.1	7	2,359
Category 2 Households (hire in labour)	1,123	802.2	152	5,905
Category 3 Households (autarchic)	681	294.6	145	2,543
Average real female wage rate (kg. of paddy/day)	4.19	0.74	2.63	7.43
Average real male wage rate (ˮ)	4.51	1.06	2.65	8.19
Age	34.9	12.9	15	60
Family size	8.56	5.2	1	28
Number of adult male workers in household	2.32	1.37	1	7
Number of adult female workers in household	1.92	1.27	1	6
Total operated farm area, hectares (all categories)	2.52	3.01	0	33
Category 1 Households (hire out labour)	0.88	1.39	0	9
Category 2 Households (hire in labour)	4.67	3.79	1.47	33
Category 3 Households (autarchic)	2.98	2.47	1.2	13
Female Workers Sample (N = 2,288)				
Total labour supply in days [a] (all categories)	56.5	35.4	0	180
Category 1 Households (hire out) (N1 = 1,103)	68.0	38.8	0	174
Category 2 Households (hire in) (N2 = 724)	51.4	28.8	0	180
Category 3 Households (autarchic) (N3 = 461)	45.5	31.1	2	170
Per capita real non-labour income [b]	575	494.2	7	3,769
Category 1 Households (hire out labour)	252	223.1	7	1,448
Category 2 Households (hire in labour)	994	543.7	76	3,769
Category 3 Households (autarchic)	729	389.1	108	2,542
Average real female wage rate (kg. of paddy/day)	4.22	0.72	2.63	7.43
Average real male wage rate (ˮ)	4.57	1.07	2.65	8.19
Age	31.9	11.7	15	60
Family size	8.64	5.4	1	28
Number of adult male workers per household	2.21	1.31	0	7
Number of adult female workers per household	2.02	1.49	1	6
Number of children aged 0 to 5	2.06	1.77	0	9
Economically Inactive Women Sample (N = 481)				
Per capita real non-labour income [b]	704.1	931	90	5,905
Family size	8.56	5.2	1	28
Number of adult male workers per household	2.03	1.25	1	7
Number of adult female workers per household	2.69	1.39	1	10
Age	34.1	14.9	15	60

[a] This is the average number of workdays in each of two six-monthly cropping seasons.
[b] Nominal per capita income (in Rupees) deflated by the village price of paddy per kg.

For the empirical implementation of Equation 7.1 a linear labour supply function is specified. The total labour days worked by an individual is regressed on the real effective wage rate, real per capita non-labour income, and a set of individual and household level characteristics. The linear labour supply model is chosen partly for the sake of simplicity and partly to aid in model identification. Because of the multiplicative form in which the θ factor affects the effective wage rate, a log-linear form cannot be used since the wage adjustment is converted into an intercept effect.

The labour supply estimation is done separately for the economically active male and female household members, allowing for cross wage effects. The general estimating equation is:

$$(7.2) \quad Ls_{ikh} = \alpha + \beta_1 w^{*m}_{kh} + \beta_2 w^{*f}_{kh} + \gamma PNLY_h + \tau S + e_{ikh}$$

$$Ls_{ikh} \quad = \text{total workdays reported for an economically active}$$
family member i in household h for crop season k ($k =$ summer or winter);[12]

$PNLY_h =$ real per capita non-labour income for household h, including the value of farm profit;

$S \qquad =$ a vector of individual and household characteristics, i.e. family size and composition, age and education level of the individual, family relationship dummies, etc.;

w^{*m}_{kh}, w^{*f}_{kh} are effective real male and female wage rates for household h in season k, which depend on the labour market exposure of the household as follows (suppressing the kh subscripts):

$w^{*j} \quad = w^j$ if household is a net seller of labour $\qquad$ (Category 1)

$w^{*j} \quad = w^j/\theta$ if household is a net buyer of labour (Category 2)

$w^{*j} \quad = w^j/\lambda^j_q$ for autarchic households $\qquad$ (Category 3)

(with $\theta \leq \lambda^j_q \leq 1$)

where $\quad j = m$ (for male) or f (for female),

q indexes the number of autarchic sample households;

and α, β_1, β_2, γ, and τ are elements of the Φ parameter set.

The effective wage rates also enter into the computation of non-labour income since the imputed cost of family labour applied to farm production must be deducted from gross farm profits.[13]

The market wage rate for hired agricultural labour computed from the MPHBS can be distinguished along three different dimensions: (i) by sample household, (ii) by gender, and (iii) by each of two six-monthly cropping seasons (the summer and winter cycle) for which farm management data was collected separately. This structure for the wage data allows some variation even in market wage rates for households within a single sample village cluster. All households in a specific sample cluster are not imputed a common average village-specific wage. Where a particular household reports hiring labour for crop production in a specific season, the market wage rate for that household is computed as the average of the actual wages paid out during that cropping season. A village level average wage is then computed as the unweighted average of the hiring in wage rate reported by all such households in the sample village. This village level average wage paid to hired labour is then imputed as the wage received by individuals in the small farm and landless households for whom no household level hired wage is reported directly in the MPHBS data.[14] All family members of a specific household in each gender group are assigned the same gender-specific wage rate since individual level wage rates are not reported in the MPHBS.

Once the nominal household level wage rates (for male and female members) are computed in this manner, they are deflated by a village-specific price of paddy. For simplicity, the real wage conversion is based only on the price of paddy rather than the price of a composite consumption good that would require complex indexing procedures.[15]

The real market wage rates (w/p) derived in this manner are converted into real effective wage rates ($w*/p$) based on the categorization of households by their net labour supply position. The derivation of a household's net labour supply position (i.e. the allocation into Category 1, 2, or 3) is based on data for the whole annual survey period and does not vary by cropping cycle. The Category 1 to 3 description also carries over to all individuals in the same household.[16] Apart from the gender distinction, all individuals in a household are assumed to face the same market and effective wage. Since all individuals of a particular gender group are assumed to be perfect substitutes for each other in own farm production, and individuals across gender groups are also perfect substitutes with a productivity difference related to the ratio of their market wage rates, it should not matter which particular family member works on the farm and which one works on the hired labour market.

Given a prior estimate of θ and the reported market wage rates, the effective wages are well defined for households in Categories 1 and 2. But the effective wage rate for autarchic households in Category 3 is not defined because λ^i_q in Eq. 7.2 cannot be identified since it varies across households. It is not appropriate either to use the market wage rate, w, adjusted by some constant factor because small changes in w will have no effect on the labour supply of individuals in autarchic households. If the market wage rate is attributed to the labour supply equilibrium of autarchic households, the expected values of the β wage coefficients are zero since $\dfrac{\partial Ls}{\partial w}^j = 0$ for individuals in Category 3 households (j = male or female).

Section 7.5 below presents some preliminary regression results based on a sample that includes the autarchic households to detect whether this condition is met. In subsequent regressions however the observations on the autarchic households are dropped.

Alternative Model Specifications

Focussing on individuals in Categories 1 and 3 only, four separate models can be defined. These differ on whether the common observed market wage or the varying effective wage rate is used; and whether or not wage slope and intercept differences are allowed for the two household categories. These alternative models can be presented in the framework of Equation 7.2 as follows (ignoring for expositional purposes only the subscripts, and the cross wage terms for male and female labour supply):

Model A: common market wage; common parameters

(7.3A) $Ls = \alpha + \beta w + \gamma PNLY + ...$ for Category 1 & 2

Model B: common market wage; varying parameters

(7.3B.1) $Ls = \alpha^1 + \beta^1 w + \gamma^1 PNLY(w) + ...$ for Category 1
(7.3B.2) $Ls = \alpha^2 + \beta^2 w + \gamma^2 PNLY(w) + ...$ for Category 2

Model C: different effective wage rates; common parameters

(7.3C.1) $Ls = \alpha + \beta w + \gamma PNLY(w/\theta) + ...$ for Category 1
(7.3C.2) $Ls = \alpha + \beta (w/\theta) + \gamma PNLY(w/\theta) + ...$ for Category 2

Model D: different effective wage rates; varying parameters

(7.3D.1) $Ls = \alpha^1 + \beta^1 w + \gamma^1 PNLY(w/\theta) + ...$ for Category 1
(7.3D.2) $Ls = \alpha^2 + \beta^2 (w/\theta) + \gamma^2 PNLY(w/\theta) + ...$ for Category 2

Note: PNLY(w) is PNLY derived from valuing the cost of family labour at wage w; and PNLY(w/θ) is derived using the effective wage rate w/θ.

All four model specifications above can be represented within the general framework of Equation 7.2 through the use of appropriate dummy variables for the intercept and slope terms. For Models B, C and D several different variants occur within each set, depending on the combinations of dummy variables used. For instance in Model C, while the wage slope parameter β is assumed to be constant across individuals in Category 1 and 2 households, the γ or the α parameter for the intercept may be allowed to differ for individuals in Category 1 and 2 households.

Within the general structure specified in Equations 7.3, Models A and B are nested; and so are C and D. The selection of the appropriate model within each choice set can be based on testing the coefficients on the appropriate dummy variables that give rise to differences in the parameter estimates for Categories 1 and 2. The main interest in this Chapter, however, is to verify whether the labour supply specification based on varying effective wage rates (Models C and D) is preferred to the specification based on the unadjusted market wage rate (Models A and B) for both household categories. These sets are not nested within each other. They differ in that the value of some of the right hand side variables are different for a subset of the observations. This can be treated as a difference in model specification when the equation is estimated over the full sample of individual members belonging to both categories.

Given the multiplicative form of the effective wage (w/θ) for Category 2 households, the version of Model C that allows the intercept (α term) to vary for Category 1 and 2 households would be indistinguishable from Model B, were it not for the fact that PNLY is computed in a different manner in Model C (and D) than in Model B. Otherwise, the β coefficient for the wage variable for Category 2 households in Model C would differ from the β^2 coefficient for Category 2 households in Model B by the constant θ factor.[17] Therefore it would not be possible to identify whether Equation 7.3B.2 or Equation 7.3C.2 had been estimated.

The fact that model identification depends critically on the values of PNLY being different in these two specifications is a potential problem. The differences in NLY computed on the basis of w or w/θ are likely to be minor. This problem is further compounded if PNLY turns out to be correlated with the error term of Equation 7.3 (which can be checked, for instance, with the Wu-Hausman test). It would then be necessary to use an instrumented version of PNLY in which case the statistical differences between the instrumented versions of PNLY(w) and PNLY(w/θ) could be even smaller.

7.5 Labour Supply Regression Results: Male Family Members

Table 7.3 summarizes the preliminary regression results for male labour supply using the observed *market wage rate* for the *full sample* of individuals in all three household categories – net sellers, net buyers and autarchic households. The results are presented for Model A (with common parameters for all households) and two versions of Model B that allow for different intercepts and slope dummies for the wage and non-labour income variables. Model B1 has an intercept dummy only, while Model B2 has dummies for the intercept and for the slopes of the own wage and non-labour income terms. The regression models of Table 7.3 include other variables for which the estimated coefficients and standard errors are not reported in that table. [18] (The full regression results for all variables in the estimating equation for several model specifications, including Model B2, are given in the Appendix Tables to this chapter).

Although the parameter estimates in Model A and B1 appear reasonable with positive own wage effects and negative income effects on labour supply, Model B2, which allows for the full set of slope and intercept dummies, is clearly superior to Models A and B1. Since Models A B1 and B2 are nested within each other the preferred specification can be determined by verifying the significance of the extra dummy variables that appear in Models B1 and B2. The full results for Model B2 (given in Appendix Table 7A.2) show that five out of the six intercept and slope dummies are highly significant, leading to significantly different wage and income effects for the three household categories. The model fit also improves considerably as one allows for additional dummy intercept and slope terms. The adjusted R-square increases from 0.29 for Model A to 0.41 for Model B2.

In Model B2 the own wage coefficients are of the expected sign (positive) and significantly different from zero for Category 1 and 2 households. The own wage coefficient is not significantly different from zero for autarchic households. This last result is as expected.[19] There is a similar difference in the income effects for the three household categories. The coefficients on the non-labour income variable are significantly negative for Categories 1 and 2, which is consistent with leisure being a normal good. The income effect on labour supply for autarchic households however has a positive sign, although the coefficient is not significantly different from zero. This discrepancy in the income effect of autarchic households is not a theoretically expected result. It does, however, add to the inference based on the own wage effects discussed above, suggesting

Table 7.3 Male Labour Supply Regressions with Common Market Wage Rates

Dependant variable:	Total days of work in cropping season			
Data subset:	Males, All Categories (1,2,3), Sample N = 2,542			
Estimation method:	OLS (heteroskedasticity consistent errors)			

	Model A common parameters		Model B1 intercept dummies only	
Variable	*coeffic.*	*t-ratio*	*coeffic.*	*t-ratio*
Male wage rate	2.777	1.53	5.505	3.50
(hire out) Category 1 (N1 = 1214)				
(hire in) Category 2 (N2 = 813)				
(autarchic) Category 3 (N3 = 515)				
Female wage rate	-4.002	-1.94	-4.034	-2.16
Non-labour income (x 100)	-1.631	-10.26	-0.621	4.31
Category 1				
Category 2				
Category 3				
Family size	0.885	2.45	1.331	3.93
Number male workers	-10.003	-9.66	-10.148	-10.56
Number female workers	0.612	0.61	0.104	0.15
Age	1.343	3.37	1.125	2.97
Age squared	-0.025	-5.26	-0.021	-4.66
Education (years)	-1.145	-2.29	-0.325	-0.07
Household head dummy	7.023	1.66	3.271	0.82
Season dummy	-4.194	-3.02	-4.166	-3.22
Intercept	108.2	11.45		
Category 1			106.4	11.98
Category 2			80.8	9.04
Category 3			68.8	7.50
Adjusted R-Square		0.29		0.39
Standard error of the estimate		35.20		32.73
Breusch-Pagan Heteroskedasticity Test	(df 19)	* 44.60	(df 21)	* 68.28
RESET (2) Test	(df 1, 2521)	* 64.09	(df 1, 2519)	* 5.09
Wu-Hausman Test for Exogeneity of:				
Non-labour income				
Non-labour income and wage rates				

Note: The parameter estimates noted above are incomplete. These regressions include several other household relationships, ethnicity/caste and regional dummy variables.
* indicates the relevant diagnostic test statistic is significant at the 5% level (or lower).
(The significance of the *t-ratio* for the parameter estimates is not indicated).

Table 7.3 *(continued)* **Male Labour Supply Regressions with Common Market Wage Rates**

Dependant variable:	Total days of work in cropping season		
Data subset:	Males, All Categories (1,2,3)		
Estimation method:	OLS (heteroskedasticity consistent errors)		

	Model B2 intercept & slope dummies	
Variable	*coeffic.*	*t-ratio*
Male wage rate		
(hire out) Category 1 (N1 = 1214)	6.301	3.21
(hire in) Category 2 (N2 = 813)	9.869	5.05
(autarchic) Category 3 (N3 = 515)	0.599	0.38
Female wage rate	-4.529	-2.37
Non-labour Income (x 100)		
Category 1	-3.307	-6.88
Category 2	-0.369	-2.52
Category 3	0.439	1.01
Family size	1.313	4.00
Number male workers	-9.728	-10.31
Number female workers	0.488	0.53
Age	1.052	2.81
Age squared	-0.021	-4.53
Education (years)	-0.360	-0.79
Household head dummy	4.058	1.01
Season dummy	-4.162	-3.27
Intercept		
Category 1	110.5	11.56
Category 2	60.1	6.24
Category 3	87.6	7.94
Adjusted R-Square		0.41
Standard error of the estimate		32.21
Breusch-Pagan Heteroskedasticity Test	(df 25)	* 148.8
RESET (2) Test	(df 1, 2515)	* 6.18
Wu-Hausman Test for Exogeneity of:		
Non-labour income	(df 26)	18.60
Non-labour income and wage rates	(df 26)	27.22

Note: The parameter estimates noted above are incomplete. These regressions include several other household relationships, ethnicity/caste and regional dummy variables.
* indicates the relevant diagnostic test statistic is significant at the 5% level (or lower).
(The significance of the *t-ratio* for the parameter estimates is not indicated).

that the labour supply behaviour of the male workers in autarchic households appears to be quite distinct from that of individuals in households that report some labour selling or buying.

The non-labour income variable used in the regressions of Table 7.3 includes farm profit derived from valuing all family labour at the market wage rate. The Wu-Hausman test for the exogeneity of non-labour income (as well as the wage rate) is reported in Table 7.3 for Model B2, which is the preferred model. The test statistic is not significant, indicating these variables are not correlated with the error term.[20] Hence, instrumental variable techniques need not be used for the non-labour income variable computed on the basis of the observed market wage rates. This is a useful result that facilitates identification between Models B and C subsequently.

The coefficients on other regression variables do not differ greatly between the three models in Table 7.3. The observed relationships with these other variables are plausible. Individual labour supply increases with age but at a decreasing rate (implied by the negative coefficient on the age-squared term). The coefficient on years of education is negative in Model B2 but it is not significantly different from zero. The cropping season dummy is significant, with a slightly smaller labour supply intercept for the winter cropping cycle. The (male) household head supplies extra labour compared to other male family members. Controlling for family size, a larger number of available male family workers reduces the labour supply of each individual male worker. But a similar relationship does not occur with respect to the number of female workers in the household. This is an indication of work sharing within gender groups.

In spite of the reasonable parameter estimates in Table 7.3, the RESET specification test indicates substantial functional mis-specification for all three models that are based on the observed market wage rate. The mis-specification error is the largest for Model A which imposes common parameters for all household categories.

Table 7.4 provides a comparison of the regression results for alternative model specifications based on assigning different effective wage rates for Category 2 households. The sub-sample of individuals from autarchic households is dropped for these regressions because for autarchic households the effective wage rate is not identified in terms of the θ parameter.[21] Model B in Table 7.4 is the same specification as Model B2 in Table 7.1 with the autarchic observations dropped. Models C and D use the w/θ effective wage for individuals in labour hiring households. All three specifications allow intercept and non-labour income slope dummies since these were significant effects in the preliminary results of Table 7.3.

Table 7.4 Male Labour Supply Regressions with Alternative Effective Wage Rates

Dependant variable:	Total days of work in cropping season
Data subset:	Males, Categories 1 & 2 only, Sample N = 2,037
Estimation method:	OLS (heteroskedasticity & 2-step consistent errors)

	Model B common market wage; intercept & wage slope dummies		Model C varying effective wage; intercept dummy only	
Variable	*coeffic.*	*t-ratio*	*coeffic.*	*t-ratio*
Male wage rate			5.687	3.52
(hire out) Category 1 (N1 = 1214)	8.861	3.75		
(hire in) Category 2 (N2 = 813)	12.469	5.25		
Female wage rate	-8.556	-3.46	-1.691	-1.05
Non-labour income (x 100)				
Category 1	-3.435	-5.82	-3.125	-6.30
Category 2	-0.468	-3.06	-0.448	-2.98
Family size	0.770	1.89	1.044	2.76
Number male workers	-10.289	-8.80	-9.433	-8.58
Number female workers	1.622	1.47	1.896	1.87
Age	1.658	3.74	1.398	3.32
Age squared	-0.028	-5.32	-0.023	-4.59
Education (years)	-0.325	-0.61	0.065	0.13
Household head dummy	7.023	1.66	4.033	0.90
Season dummy	-4.062	-2.75	-4.334	-3.11
Intercept				
Category 1	110.1	9.72	94.1	9.59
Category 2	61.2	5.98	47.1	4.42
Adjusted R-Square		0.32		0.39
Standard error of the estimate		33.35		31.53
Breusch-Pagan Heteroskedasticity Test	(df 22)	* 95.17	(df 21)	* 44.03
RESET (2) Test	(df 1, 2003)	2.17	(df 1, 2004)	2.03
Model Selection Diagnostics:				
Akaike Final Prediction Error		1125.3		1005.7
Schwartz Criteria		1199.3		1071.8

Note: The parameter estimates given above are incomplete. These regressions include several other household relationships, ethnicity/caste and regional dummy variables.
The *t-ratio* reported for Models C and D are based on the adjusted variance–covariance matrix for the two step estimator as described in Section 4.4.
* indicates the relevant diagnostic test statistic is significant at the 5% level (or lower).
(The significance of the *t-ratios* for the parameter estimates is not indicated).

Table 7.4 *(continued)* Male Labour Supply Regressions with Alternative Effective Wage Rates

Dependant variable:	Total days of work in cropping season	
Data subset:	Males, Categories 1 & 2 only (excluding autarchic)	
Estimation method:	OLS (heteroskedasticity & 2-step consistent errors)	

	Model D varying effective wage; intercept & wage slope dummies	
Variable	*coeffic.*	*t-ratio*
Male wage rate		
(hire out) Category 1 (N1 = 1214)	5.796	3.06
(hire in) Category 2 (N2 = 813)	5.667	5.05
Female wage rate	-1.720	-1.06
Non-labour income (x 100)		
Category 1	-3.217	-6.27
Category 2	-0.448	-2.97
Family size	1.043	2.76
Number male workers	-9.428	-8.56
Number female workers	1.897	1.87
Age	1.398	3.32
Age squared	-0.023	-4.59
Education (years)	0.063	0.12
Household head dummy	4.029	0.90
Season dummy	-4.339	-3.11
Intercept		
Category 1	94.0	9.80
Category 2	46.7	4.06
Adjusted R-Square		0.39
Standard error of the estimate		31.53
Breusch-Pagan Heteroskedasticity Test	(df 22)	* 44.09
RESET (2) Test	(df 1, 2003)	2.04
Model Selection Diagnostics:		
Akaike Final Prediction Error		1006.7
Schwartz Criteria		1075.8

Note: The parameter estimates given above are incomplete. This regression includes several other household relationships, ethnicity/caste and regional dummy variables.
The *t-ratio* reported for Models C and D are based on the adjusted variance–covariance matrix for the two step estimator as described in Section 4.4.
* indicates the relevant diagnostic test statistic is significant at the 5% level (or lower).
(The significance of the *t-ratios* for the parameter estimates is not indicated).

With this specification, the only difference between the three models presented in Table 7.4 relates to the definition of the effective wage rate, and whether or not own wage slope dummies are included as additional variables. Note that as in Table 7.3, the results in Table 7.4 do not report the estimate and standard errors for the actual intercept and slope dummy coefficients. These dummy variable coefficients have been added to the estimated base category parameters to compute the actual parameter value for each category reflected by the dummy variables. [22]

In comparing Models C and D, which are nested, the own wage coefficients for labour selling and labour hiring households are not significantly different. Once the higher effective wage rate is allowed for Category 2 households, there is no significant difference in the own wage slope parameter for individuals in Category 1 and 2 households. [23] Hence, Model D reduces to Model C.

In comparing Model C with Model B (where the latter uses the observed market wage rate with varying own wage slopes), Model C is the preferred specification on the basis of several diagnostic statistics – higher adjusted R-square, smaller Akaike Prediction Error, and on the basis of the *J* test for non-nested models reported in Table 7.5. [24]   In the preferred Model C, the Wu-Hausman test for exogeneity of the non-labour income variable based on the effective wage rate – $PNLY(w/\theta)$ – is insignificant again. [25]

Table 7.5 Male Labour Supply: *J* tests between Model B and C [26]

	Additional variable	Estimated coefficient	Stand. error	*t* statistic	Inference
Model B extended	predicted values from Model C	1.38	0.33	4.16	Reject Model B in favour of C
Model C extended	predicted values from Model B	0.56	0.42	1.35	Do not reject C in favour of B

Although Model C is clearly preferred on statistical grounds, the comparison of the income and wage elasticities based on Model C and Model B, given in Table 7.6, do not lead to very striking differences, with one exception. The cross wage elasticity of male labour supply with respect to the female wage rate is insignificant in Model C while it is significantly negative in Model B. Table 7.6 also presents the elasticity values computed from a Model C specification that drops the female real wage variable from the male labour supply equation. This version of Model C restricts the cross wage elasticity of male labour supply with respect to the female wage rate to be zero.

The results for these two alternative specifications of Model C are similar with a slightly higher own wage effect for the specification that allows for a negative cross wage effect.[27] The own wage elasticities in Model B give slightly higher estimates than those based on either version of Model C. The estimates for the income elasticities are very similar in all model specification and for both categories of households. The income elasticity values are in all cases quite low – less than 0.1 in absolute value.

Table 7.6 shows significant differences in the computed wage elasticities at the mean of the data for male workers in Category 1 and 2 households. The (positive) uncompensated own wage elasticities are substantially higher for workers in Category 2 households than for those in Category 1 households in all three specifications. This result of a higher elasticity for workers in Category 2 households (which hire in labour) is consistent with the theoretical prediction that a backward bending labour supply curve (negative uncompensated own wage elasticities) is possible only for net sellers of labour in Category 1 households (Strauss, 1986: 76).

The complete regression results for Model C of male labour supply, with and without the female cross wage variable, are presented in Appendix Table 7A.3 with the two step error-correction procedure to account for the fact that θ is a pre-estimated parameter. In addition to the expected signs on the wage and income variables, the effects of the other variables are also reasonable. Holding family size constant, a higher number of male workers in the family reduces the workdays of an individual worker; but there is no corresponding effect from the number of female workers in the family. Workdays increase with age but at a decreasing rate. The diagnostic statistics indicate that in spite of the inflexible responses inherent in a linear labour supply equation, the RESET test does not detect functional form mis-specification in Model C (nor B).[28]

In summary, the specification of a labour supply function for male family members that is consistent with the prior estimate of the higher efficiency of family labour in own farm production performs better than a model which assumes a common wage and constant parameters for workers from all households.

Table 7.6 Estimated Wage and Income Elasticities of Male Labour Supply *(with standard errors below in italics)*

Household type	Elasticity with respect to:			Compensated own wage elasticity
	own wage	income	female wage	
Model B : common market wage, and varying slope parameters				
(hire out) Category 1	0.355	-0.078	-0.323	0.644
	0.095	*0.013*	*0.093*	*0.109*
(hire in) Category 2	0.675	-0.067	-0.438	0.671
	0.128	*0.0219*	*0.121*	*0.128*
Model C : varying effective wage, no wage slope dummies version 1				
Category 1	0.228	-0.070	-0.064 #	0.488
	0.065	*0.011*	*0.061*	*0.078*
Category 2	0.410	-0.061	-0.113 #	0.404
	0.116	*0.203*	*0.101*	*0.116*
Model C : varying effective wage, no wage slope dummies, *no cross wage effect* version 2				
Category 1	0.181	-0.071		0.443
	0.057	*0.011*		0.074
Category 2	0.326	-0.061		0.320
	0.102	*0.021*		*0.102*

Note: Computed at the mean of the data excluding individuals in autarchic households.
# denotes not significantly different from zero.
Category 1 households are net sellers of labour, and Category 2 are net buyers of labour.

7.6 Labour Supply Regression Results: Female Family Members

This section presents the results of the labour supply regression for female family members of the sample households. The estimation of the female labour supply regressions has an additional step to correct for the sample selection bias that may result from the large proportion of individual female family members who are reported to be economically inactive in the MPHBS employment data.

Out of the 1,688 female family members enumerated in the sample households, almost 30% (481) individuals are economically inactive, according to the definitions adopted in the MPHBS.[29] This is a significant proportion. It is well known that sample selection bias may occur when the labour supply equations are estimated from the sample of the economically active individuals only (Heckman, 1979 & 1980). To correct for this possibility, Heckman's two step estimation procedure (Heckit) is used for the labour supply regressions for female family members.

In the first step of the Heckit procedure, a probit model for labour force participation is estimated using the full sample of economically active and inactive female individuals. In the second step the labour supply (workdays) equation is estimated for the sample of economically active women only, but with an added explanatory variable (the inverse Mill's ratio) which controls for the possible correlation of the error terms between the participation and labour supply equations.[30]

7.6.1 Probit Labour Force Participation Model

The maximum likelihood probit estimate of the female labour force participation equation is given in Table 7.7. The binary dependent variable is defined so that those who are economically active are assigned the value of one. Therefore a positive coefficient on the regression parameters means an increase in the probability of labour market participation. Since the estimated probit coefficients do not directly give the marginal effects of a change in the explanatory variables, these are computed separately in Table 7.7 at the mean of the data, together with the implied elasticities.[31]

A proper implementation of the Heckit procedure requires that some of the explanatory variables in the probit equation be unrelated to the variables in the second stage labour workdays equation (Heckman, 1980). While a completely independent set of instruments is not available, two separate adjustments are made. The non-labour income in the probit equation is defined in a different manner. In addition, since the Heckit two step method already includes a special procedure for adjustments in the variance-covariance matrix for the parameters of the labour supply equation in the second step,[32] the labour force participation equation is modelled to be independent of the θ parameter estimated in the production function. The wage rate variable used in Table 7.7 is the observed market wage. A dummy variable category for households that use hired labour is created. Similarly the non-labour income variable excludes farm profit so that an effective wage valuation of family labour need not be made at this step. A proxy for the imputed farm profit component of non-labour income is gross harvested area, with a dummy variable to control for households that are tenants. The returns to farm operation will be substantially less for tenant households than for owner cultivators due to land rental payments.

The probit regression results indicate that a higher market wage rate, a higher non-labour income (excluding farm profit) and a larger farm size increase the probability of labour market participation for female family members. The squared term of area harvested in significantly negative, implying the effect of increasing farm size will eventually be negative. Age has a similar non-linear effect on labour market participation, with the age term being positive and the age squared term being negative. Other demographic variables also have plausible effects. The coefficients on family size and the number of male workers are positive, but not significant. A larger number of adult female members in a household, however, reduces the probability of labour market participation for a particular individual. This implies unequal work sharing among female members in a household. The presence of children aged 0-5 in a household reduces the probability of female labour market participation, but there is no statistically significant effect of the presence of children aged 6 to 9. The dummy variable for tenant households is significantly positive. The coefficient on the dummy for hiring in labour is negative but insignificant. Several of the regional dummies are also significant.

The model fit statistics indicate a reasonable fit for a binary dependant variable model. The McFadden R-square is 0.2. The prediction success table shows that while the overall percentage of right predictions is high (at 78%), the bulk of the right predictions are for individuals who are economically active. Among the 481 individuals who are not economically active, the number of right predictions is about 41%. Most of the prediction errors result from not being able to correctly predict non-participation for specific individuals. Nevertheless, given that most of the explanatory variables used in the probit regression are household level variables, while the data show differences in the labour force participation among female members of the same household, the probit model of Table 7.7 performs reasonably well.

7.6.2 Labour Supply (Days Worked) Models

The labour supply regressions for female workers are also carried out only for the sample of individuals in Category 1 and Category 2 households. Observations from the autarchic households are dropped from the final

Table 7.7 Female Labour Force Participation Equation

Data subset:	All Female active and inactive persons (N = 1688)				
Estimation method:	Probit Maximum Likelihood				

Variable	estimated coefficient	asymptotic stand. error	t-ratio	marginal effect at means	marginal elasticity at means
Female real wage rate	0.1154	0.055	2.11	0.036	0.200
Family size	0.0236	0.015	1.57	0.007	0.086
Number male workers	0.0410	0.038	1.07	0.013	0.036
Number female workers	-0.1639	0.039	-4.22	-0.051	-0.173
Age	0.1029	0.013	7.82	0.032	1.378
Age squared	-0.0015	0.000	-8.47	-0.001	-0.741
Non-labour household income (excludes imputed farm profit)	0.0022	0.001	2.35	0.001	0.048
Farm area harvested	0.4081	0.223	1.83	0.000	0.067
Farm area harvested squared	-0.0004	0.000	-3.99	0.000	-0.058
Intercept	-0.7949	0.347	-2.29	-0.248	-0.327
Dummy Variable Categories:					
Western region	0.6405	0.080	7.96	0.200	0.055
Far-western region	-0.7312	0.140	-5.23	-0.228	-0.045
Household head	0.2909	0.183	1.59	0.091	0.003
Unmarried dummy	0.1651	0.203	0.81	0.052	0.004
Labour hired in dummy	-0.2377	0.363	-0.66	-0.074	-0.037
Presence of children aged 0-5	-0.1169	0.081	-1.44	-0.037	-0.039
Presence of children aged 6-9	0.0094	0.066	0.14	0.003	0.002
Tenant household	0.3572	0.114	3.14	0.112	0.016

Log Likelihood function	=	-1369.5
Log Likelihood (0)	=	-1727.9
Likelihood Ratio Test	=	* 716.8 χ^2 with 24 df
Maddala R-Square		0.220
Cragg-Uhler R-Square		0.315
McFadden R-Square		0.207
(adjusted for degrees of freedom)		0.199

PREDICTION SUCCESS TABLE

		ACTUAL	
		0	1
PREDICTED	0	199	86
	1	282	1121
All Cases		481	1207
Number of Right Predictions	=	1320	
Proportion of Right Predictions	=	0.78	

Note: The probit model parameter estimates given above are incomplete.

regression because the effective wage rate is not observed for such individuals. Using the market wage rate for individual female workers in autarchic households leads to anomalous results, just as in the case of the labour supply regression for male household members.[33]

Another adjustment made in the female labour supply regressions is to drop the cross wage term – the male wage rate variable. With both wage rates included in the regressions, neither becomes significant in most specifications. Since in the final specification of Model C for male workers the female cross wage term was insignificant, it is also theoretically consistent to drop the male wage rate variable from the female labour supply regressions.[34]

The labour supply regression results for female family members are presented in summary form in Tables 7.8 to 7.10 for model specifications B, C and D, respectively. Under each model specification there are two sets of parameter estimates: one based on OLS, ignoring the sample selection problem; the other is based on the second step of the Heckit procedure with the inverse Mill's ratio (IMR) as an additional variable.

Model B equates the opportunity cost of family labour to the observed market wage rate for female labourers, but allows for intercept and slope dummies for individuals in Category 1 and 2 households. The standard errors of the coefficients reported for Model B are White's heteroskedasticity consistent errors for the OLS estimates, since heteroskedasticity is indicated in all model specifications. For the Heckit estimates of Model B, the reported standard errors reflect the standard adjustment required when the IMR is included as an additional variable in the regression because the IMR is also a generated regressor.[35] The Heckit standard errors do not explicitly control for heteroskedasticity. Models C and D use the effective wage rate variable based on the w/θ adjustment for Category 2 households. Hence, in Tables 7.9 and 7.10 an additional set of corrected standard errors is reported to account for the fact that θ is an estimated parameter from the production function equation.[36]

The parameter estimates of all specifications are quite similar. The own wage effect on female labour supply is significantly positive, and the non-labour income effect is significantly negative, as was the case with the male labour supply results. Both the intercept dummy and the slope dummy on non-labour income for Category 2 households are always significant.[37]

In all specifications the IMR variable is insignificant. The estimated correlation between the errors in the labour force participation equation and the labour supply equation is around 0.3. The Heckit model, however, is still meaningful even though the coefficient on IMR is not significant.

Table 7.8.1 Female Labour Supply Regression: Model B (OLS)

Dependant variable:	Total days of work in cropping season
Data subset:	Females, Categories 1and 2 only (N = 1,827)
	(excluding autarchic and inactive persons)
Estimation method:	OLS (heteroskedasticity consistent errors)
Model description	Common market wage;
	with intercept & slope dummies

Variable	estimated coefficient	standard error	t-ratio
Female wage rate			
(hire out) Category 1	4.194	1.828	2.29
(hire in) Category 2	8.061	1.670	4.83
Non-labour income (x 100)			
Category 1	-2.289	0.511	-4.48
Category 2	-0.318	0.219	-1.45
Family size	2.810	1.123	2.50
Number male workers	-7.265	1.385	-5.25
Number female workers	-3.877	1.468	-2.64
Number of children aged 0-5	-3.463	1.247	-2.78
Age	1.316	0.393	3.35
Age squared	-0.024	0.005	-4.58
Education (years)	-1.732	1.020	-1.70
Household head dummy	13.891	4.748	2.93
Season dummy	-6.627	1.377	-4.81
Inverse Mill's Ratio (x 100)			
Intercept			
Category 1	45.0	10.4	4.33
Category 2	35.4	9.3	3.81

Adjusted R-Square	0.30	
Standard error of the estimate	29.68	
Log Likelihood	-8766.9	
Breusch-Pagan Heteroskedasticity Test	* 166.7	$\chi^2$with 23 df
RESET (2) Test	2.90	F with 1, 1802 df
Model Selection Diagnostics:		
Akaike Final Prediction Error	895.4	
Schwartz Criteria	980.2	
Correlation between participation and labour supply equations (ρ)		

Note: The parameter estimates given above are incomplete. This regression includes several other household relationships, ethnicity/caste and regional dummy variables.
* indicates the relevant diagnostic test statistic is significant at the 5% level (or lower).
(The significance of the *t-ratios* for the parameter estimates is not indicated).

Table 7.8.2 Female Labour Supply Regression: Model B (Heckit)

Dependant variable:	Total days of work in cropping season
Data subset:	Females, Categories 1and 2 only (N = 1,827)
	(excluding autarchic and inactive persons)
Estimation method:	Heckit (with standard errors adjusted for IMR)
Model description	Common market wage;
	with intercept & slope dummies

Variable	estimated coefficient	standard error	t-ratio
Female wage rate			
(hire out) Category 1	4.742	1.794	2.64
(hire in) Category 2	8.016	1.670	4.80
Non-labour income (x 100)			
Category 1	-2.280	0.469	-4.87
Category 2	-0.324	0.219	-1.48
Family size	2.825	1.162	2.43
Number male workers	-7.035	1.505	-4.67
Number female workers	-4.254	1.573	-2.70
Number of children aged 0-5	-3.547	1.313	-2.70
Age	1.680	0.546	3.08
Age squared	-0.030	0.008	-3.86
Education (years)	-2.855	1.751	-1.63
Household head dummy	14.844	4.602	3.23
Season dummy	-6.632	1.377	-4.82
Inverse Mill's Ratio (x 100)	*7.931*	*8.429*	*0.94*
Intercept			
Category 1	35.7	14.3	2.50
Category 2	25.9	11.5	2.25

Adjusted R-Square	0.30	
Standard error of the estimate	29.68	
Log Likelihood	-8766.4	
Breusch-Pagan Heteroskedasticity Test	* 168.6	$\chi^2$with 24 df
RESET (2) Test	2.67	F with 1, 1801 df
Model Selection Diagnostics:		
Akaike Final Prediction Error	895.9	
Schwartz Criteria	983.8	
Correlation between participation and	0.271	
Labour supply equations (ρ)		

Note: The parameter estimates given above are incomplete. This regression includes several other household relationships, ethnicity/caste and regional dummy variables.
* indicates the relevant diagnostic test statistic is significant at the 5% level (or lower).
(The significance of the *t-ratios* for the parameter estimates is not indicated).

Table 7.9.1 Female Labour Supply Regression: Model C (OLS)

Dependant variable:	Total days of work in cropping season			
Data subset:	Females, Categories 1and 2 only (N = 1,827)			
	(excluding autarchic and inactive persons)			
Estimation method:	OLS (with adjusted standard errors)			
Model description	Varying effective wage; no wage slope dummy			
Variable	*estimated*	*hetcov*	*hetcov + θ adj.*	
	coefficient	*st. error*	*st. error*	*t- ratio*
Female wage rate	5.120	1.200	1.274	4.02
Non-labour income (x 100)				
(hire out) Category 1	-2.356	0.502	0.503	-4.69
(hire in) Category 2	-0.510	0.220	0.220	-2.32
Family size	2.852	1.122	1.122	2.54
Number male workers	-7.322	1.386	1.386	-5.28
Number female workers	-3.954	1.467	1.467	-2.70
Number of children aged 0-5	-3.590	1.245	1.247	-2.88
Age	1.333	0.393	0.393	3.39
Age squared	-0.025	0.005	0.005	-4.63
Education (years)	-1.700	1.014	1.015	-1.67
Household head dummy	13.775	4.777	4.777	2.88
Season dummy	-6.628	1.377	1.377	-4.81
Inverse Mill's Ratio (x 100)				
Intercept				
Category 1	41.6	8.9	9.1	4.59
Category 2	41.2	8.7	8.9	4.63

Adjusted R-Square	0.30	
Standard error of the estimate	29.66	
Log Likelihood	-8766.1	
Breusch-Pagan Heteroskedasticity Test	* 166.8	$\chi^2$with 22 df
RESET (2) Test	3.66	F with 1, 1803 df
Model Selection Diagnostics:		
Akaike Final Prediction Error	893.7	
Schwartz Criteria	975.4	
Correlation between participation and		
labour supply equations (ρ)		

Note: The parameter estimates given above are incomplete. The standard errors reported under the *hetcov* column are White's hetero-skedasticity consistent errors (using the Hetcov option in the OLS command in Shazam). The *hetcov + θ adj.* column adds a positive definite matrix, resulting from the two step error adjustment due to θ being pre-estimated, to the *hetcov* variance-covariance matrix.
* indicates the relevant diagnostic test statistic is significant at the 5% level (or lower).
(The significance of the *t-ratios* for the parameter estimates is not indicated).

Table 7.9.2 Female Labour Supply Regression: Model C (Heckit)

Dependant variable:	Total days of work in cropping season
Data subset:	Females, Categories 1and 2 only (N = 1,827)
	(excluding autarchic and inactive persons)
Estimation method:	Heckit (with adjusted standard errors)
Model description	Varying effective wage; no wage slope dummy

Variable	estimated coefficient	heckit st. error	heckit + θ adj. st. error	t- ratio
Female wage rate	5.374	1.257	1.370	3.92
Non-labour income (x 100)				
(hire out) Category 1	-2.326	0.465	0.466	-4.99
(hire in) Category 2	-0.516	0.256	0.256	2.01
Family size	2.885	1.160	1.160	2.49
Number male workers	-7.067	1.497	1.499	-4.71
Number female workers	-4.424	1.569	1.569	-2.82
Number of children aged 0-5	-3.695	1.311	1.313	-2.82
Age	1.764	0.528	0.529	3.33
Age squared	-0.031	0.007	0.007	-4.15
Education (years)	-3.037	1.700	1.708	-1.78
Household head dummy	14.909	4.576	4.577	3.26
Season dummy	-6.635	1.377	1.377	-4.82
Inverse Mill's Ratio (x 100)	*9.586*	*8.046*	*8.087*	*1.19*
Intercept				
Category 1	46.4	12.1	12.4	3.74
Category 2	28.3	11.7	11.9	2.38

Adjusted R-Square	0.30	
Standard error of the estimate	29.65	
Log Likelihood	-8765.4	
Breusch-Pagan Heteroskedasticity Test	* 168.7	χ^2 with 23 df
RESET (2) Test	3.26	F with 1, 1802 df
Model Selection Diagnostics:		
Akaike Final Prediction Error	894.0	
Schwartz Criteria	978.7	
Correlation between participation and labour supply equations (ρ)	0.325	

Note: The parameter estimates given above are incomplete. See Appendix Table 7A.4 for complete results. The standard errors reported under the *heckit* column are adjusted errors for the standard two step Heckit procedure. The *heckit + θ adj.* column adds a positive definite matrix, resulting from the error adjustment for the fact that θ is pre-estimated, to the *heckit* variance-covariance matrix.

* indicates the relevant diagnostic test statistic is significant at the 5% level (or lower).

(The significance of the *t-ratios* for the parameter estimates is not indicated).

Table 7.10.1 Female Labour Supply Regression: Model D (OLS)

Dependant variable:	Total days of work in cropping season
Data subset:	Females, Categories 1and 2 only (N = 1,827)
	(excluding autarchic and inactive persons)
Estimation method:	OLS (with adjusted standard errors)
Model description	Varying effective wage with wage slope dummy

Variable	estimated coefficient	hetcov st. error	hetcov + θ adj. st. error	hetcov + θ adj. t- ratio
Female wage rate				
(hire out) Category 1	4.062	1.832	1.833	2.22
(hire in) Category 2	5.889	1.251	1.557	3.78
Non-labour income (x 100)				
Category 1	-2.306	0.510	0.510	-4.52
Category 2	-0.507	0.219	0.223	-2.27
Family size	2.890	1.123	1.123	2.57
Number male workers	-7.380	1.386	1.386	-5.32
Number female workers	-3.989	1.469	1.469	-2.72
Number of children aged 0-5	-3.599	1.245	1.247	-2.89
Age	1.312	0.393	0.393	3.34
Age squared	-0.024	0.005	0.005	-4.57
Education (years)	-1.648	1.015	1.015	-1.62
Household head dummy	13.731	4.759	4.759	2.89
Season dummy	-6.629	1.377	1.377	-4.81
Inverse Mill's Ratio (x 100)				
Intercept				
Category 1	46.0	10.4	10.4	4.41
Category 2	37.6	12.0	12.0	3.14

Adjusted R-Square	0.301	
Standard error of the estimate	29.64	
Log Likelihood	-8765.7	
Breusch-Pagan Heteroskedasticity Test	* 167.8	$\chi^2$with 28 df
RESET (2) Test	3.42	F with 1, 1799 df
Model Selection Diagnostics:		
Akaike Final Prediction Error	894.2	
Schwartz Criteria	979.0	
Correlation between participation and labour supply equations (ρ)		

Note: The parameter estimates given above are incomplete. See Note to Table 7.9.1 for the description of the standard errors under the *hetcov* and the *hetcov* + θ *adj.* columns.
* indicates the relevant diagnostic test statistic is significant at the 5% level (or lower).
(The significance of the *t-ratios* for the parameter estimates is not indicated).

Table 7.10.2 Female Labour Supply Regression: Model D (Heckit)

Dependant variable:	Total days of work in cropping season
Data subset:	Females, Categories 1and 2 only (N=1,827)
	(excluding autarchic and inactive persons)
Estimation method:	Heckit (with adjusted standard errors)
Model description	Varying effective wage with wage slope dummy

Variable	estimated coefficient	heckit st. error	heckit + θ adj. st. error	t- ratio
Female wage rate				
(hire out) Category 1	4.622	1.791	1.792	2.58
(hire in) Category 2	5.853	1.497	1.722	3.40
Non-labour income (x 100)				
Category 1	-2.296	0.468	0.468	-4.91
Category 2	-0.513	0.253	0.253	-2.03
Family size	2.905	1.161	1.161	2.50
Number male workers	-7.145	1.503	1.503	-4.75
Number female workers	-4.376	1.571	1.571	-2.79
Number of children aged 0-5	-3.685	1.311	1.312	-2.81
Age	1.684	0.545	0.545	3.09
Age squared	-0.030	0.008	0.008	-3.87
Education (years)	-2.797	1.748	1.748	-1.60
Household head dummy	14.706	4.589	4.589	3.20
Season dummy	-6.634	1.377	1.377	-4.82
Inverse Mill's Ratio (x 100)	*8.121*	*8.421*	*8.422*	*0.96*
Intercept				
Category 1	36.4	14.3	14.3	2.56
Category 2	27.9	12.0	12.0	2.32

Adjusted R-Square	0.30	
Standard error of the estimate	29.64	
Log Likelihood	-8765.2	
Breusch-Pagan Heteroskedasticity Test	* 169.4	$\chi^2$with 24 df
RESET (2) Test	3.18	F with 1, 1801 df
Model Selection Diagnostics:		
Akaike Final Prediction Error	894.8	
Schwartz Criteria	982.5	
Correlation between participation and	0.276	
labour supply equations (ρ)		

Note: The parameter estimates given above are incomplete. See Note to Table 7.9.2 for
the description of the standard errors under the *heckit* and the *heckit + θ adj.* columns.
* indicates the relevant diagnostic test statistic is significant at the 5% level (or lower).
(The significance of the *t-ratios* for the parameter estimates is not indicated).

A likely reason for the insignificant coefficient on IMR is that the other regressors in the labour supply equation are similar to the regressors in the probit equation. This can lead to a high degree of multi-collinearity between the IMR variable and the other regressors in the labour supply equation, resulting in an insignificant coefficient for IMR. Because of this possibility the Heckit specification is preferred to the OLS (which drops the IMR variable). The tests for model selection below as well as the computed values of the various elasticities of female labour supply are based on the Heckit specification. Since the differences in the estimated parameters with and without the IMR variable are small, this choice of a preferred specification is not very consequential in deriving the elasticities.

As for model selection, as noted before, Models C and D are nested, but B is an independent model with a different wage variable.[38] The estimated coefficient on the additional own wage slope dummy variable for Category 2 households in Model D is insignificant in both the OLS and Heckit specifications.[39] Therefore, Model D reduces to Model C with a common wage coefficient for all individuals, but with the effective wage rate being w/θ for individuals in Category 2 households. As in the case of the male labour supply regression results, the model selection choice for the female labour supply specifications is also between Model B (with a common market wage rate and varying wage coefficients) and Model C (with an adjusted effective wage rate and common coefficients, with the implied difference in the non-labour income variable as well).

In comparing Model C with Model B, the RESET test does not reject either specification. Model C has a higher adjusted R-square and a smaller residual variance than Model B (which has one additional parameter). Because of this, Model C is preferred on the basis of the Akaike Prediction Error and the Schwartz Criteria (both of which are smaller for Model C). The non-nested *J* test also clearly shows that Model C is preferred to Model B, as indicated in Table 7.11. The full regression results for Model C for the female workers sample is given in Appendix Table 7A.4.

While Model C is the preferred specification, the resulting differences in the elasticities of female labour supply between Models B and C are fairly minor (unlike the case for the male labour supply elasticities where there were some significant differences). The estimates for the labour supply elasticities of female workers are given in Table 7.12 using the coefficients of the Heckit specification.[40] The relatively large difference observed in Model B between the two household categories for the uncompensated and compensated own wage elasticities is reduced slightly with the effective wage specification in Model C. In both specifications the

Table 7.11 Female Labour Supply: *J* tests between Model B and C

	Additional variable	Estimated coefficient	Stand. error	*t* statistic	Inference
Model B * extended	predicted values from Model C	6.91	1.02	6.77	Reject Model B in favour of C
Model C * extended	predicted values from Model B	-2.06	1.69	1.25	Do not reject C in favour of B

* Based on the Heckit specification, using adjusted standard errors.

difference in the own wage elasticity for individuals in Category 1 and Category 2 households is statistically significant, even though both elasticities are derived from the same estimated parameter.[41] The higher own wage elasticity of labour supply for individuals in labour hiring in households (Category 2) in comparison to Category 1 households (which consist mainly of landless and small farm households) is the expected theoretical result (Rosenzweig, 1980).

Comparing Table 7.12 with the corresponding elasticities of male labour supply (Table 7.6) the own wage elasticities in the preferred Model C are higher for female workers than for males. Similarly the income elasticity of labour supply is also slightly higher for female workers. A relatively higher wage and income elasticity for women than for men is an expected result commonly found in both developed and developing country settings (Killingsworth and Heckman, 1986; Bardhan, 1979).

7.7 Summary

The labour supply regression results in this Chapter represent the second step in the sequential estimation strategy of a farm household model that allows for labour heterogeneity. This step had two objectives. Firstly, to generate a complete set of the labour supply elasticities to describe, in an aggregate form, the leisure preferences of the farm households.[42] Secondly, to provide additional evidence for the labour heterogeneity results of Chapter 6 by verifying that the observed labour supply behaviour is theoretically consistent with the finding of a higher productivity of family labour when applied to own farm cultivation.

Table 7.12 Estimated Wage and Income Elasticities of Female Labour Supply *(with standard errors below in italics)*

Household type	Elasticity with respect to: own wage	income	Compensated own wage elasticity
Model B: common market wage, varying slope parameters; no cross wage effect			
(hire out) Category 1	0.280 *0.098*	-0.082 *0.018*	0.402 *0.120*
(hire in) Category 2	0.665 *0.149*	-0.077 *0.044*	0.662 *0.139*
Model C: varying effective wage; no wage slope dummies; no cross wage effect			
Category 1	0.317 *0.075*	-0.085 *0.018*	0.558 *0.141*
Category 2	0.594 *0.104*	-0.106 *0.043*	0.431 *0.082*

Note: Computed at the mean of the data, excluding individuals in autarchic households. Category 1 households are net sellers of labour, and Category 2 are net buyers of labour.

An effective labour aggregator or composite function of the linear form (Le = F + θH) implies that a higher effective wage is applicable for family labour in own farm work as a substitute for hired labour. The underlying farm household model, however, is still recursive since the wage gap is given by a constant (θ), and the effective wage rate becomes exogenous to the household's production and labour supply decisions. This model structure can be estimated through a two step strategy, allowing some of the regressors in the labour supply equation to be based on the θ parameter estimated in the production function.

An independent verification of the production function estimation result that θ < 1 can be made by comparing alternative labour supply model specifications. For both the male and female workers sample, the model allowing for labour heterogeneity (Model C) is always preferred to the model based on homogeneous labour (Model B which has a common market wage rate for all household categories). This result is verified through several diagnostic statistics on model specification, including the *J* test for non-nested models, which gives a clear verdict favouring Model C.

In the specifications based on the effective wage rate, the wage slope dummy variable for labour hiring households is not significant (in Model D). This means the effective wage rates based on the θ parameter estimated from the production function are correctly defined. If the true effective wage rates for family labour in households using hired labour were substantially different from w/θ, the wage slope dummy variable in Model D would have likely have been significant. This is additional evidence that the value of θ estimated in the production function regressions is consistent with the labour supply behaviour of farm household members.

Another implication of $\theta < 1$ is that households should not engage simultaneously in hiring in and hiring out of labour. This prediction is strongly supported. Only about 3% of the sample households report simultaneous labour hiring in and hiring out, even though the incidence of labour hiring in is not limited to big farmers.

Another important finding is that the estimated values for the wage and income elasticities of labour supply differ between the specifications based on the effective wage variable and those based on the market wage rate. The differences are more striking for the elasticities of male labour supply.

Apart from the model selection diagnostic results, the overall regression results for the model based on effective wage rates are very reasonable. The estimated parameters indicate significantly positive own wage effects and negative income effects; and female labour supply is more elastic than male labour supply with respect to both wage and income. There are also substantial differences in the wage and income elasticities of labour supply for individuals who are in households that are net buyers and net sellers of labour. These are in the theoretically predicted direction. The labour supply regression equations for autarchic households using the observed market wage rate do not give meaningful results, which is also a theoretically expected outcome.

In conclusion, a simple linear labour supply model with adjusted wage rates provides a reasonable description of the labour supply behaviour of farm household members in the *tarai* region of Nepal. The regression results for both male and female family members are consistent with the prior estimate of a higher efficiency of family labour in own farm production. These results provide independent corroboration for the labour heterogeneity indicated in the production function estimation.

Appendix 7 Complete Labour Supply Regression Results for Selected Model Specifications

This appendix presents the full set of parameter estimates and diagnostic statistics for some selected model specifications for which only summary regression results were presented in the main body of Chapter 7. It contains four tables that are as follows:

Appendix Table 7A.1 gives the definition of the actual variable names used in the complete regression results presented in the succeeding tables.

Appendix Table 7A.2 presents the full results for Model B2 of Table 7.3 in the main text of Chapter 7. This Model B2 is estimated for the sample of male workers from all three household categories – labour selling, labour hiring and autarchic – using the observed market wage rates for hired labour to represent the opportunity cost of family labour for all individuals.

Appendix Table 7A.3 presents the complete results for Model C of Table 7.4 in the main text of Chapter 7. This Model C is estimated for the sample of male workers from only the labour selling and labour hiring households (Categories 1 and 2), using the effective wage rates based on θ. Two versions of Model C are estimated, with and without the female cross wage variable.

Appendix Table 7A.4 presents the complete results for Model C estimated for the sample of female family workers from the labour selling and labour hiring households, using the effective wage rates based on θ. This table gives the full regression results that go with the Heckit specification of Model C for female labour supply (Table 7.9.2) in the main text of Chapter 7.

Appendix Table 7A.1 Variable Definitions for Appendix Regressions

AGE	Age of individual
AGESQ	Age squared
CONS	Constant term
DISDM21	Dummy variable for western region district (*Rupandehi*)
DISDM23	Dummy variable for far western region district (*Kailali*)
DMRLHH1	Dummy for household head
DMRLHH3	Dummy for own children of household head
DMRLHH4	Dummy for parents of household head
DMRLHH6	Dummy for siblings of household head
EDUCYR	Years of education of individual
ETHHLDM	Dummy for low caste status of Hill region origin
ETHTLDM	Dummy for low caste status of *Tarai* region origin
FMNOFLF2	Number of economically active women aged 15-60 in household
FMNOMLF2	Number of economically active men aged 15-60 in household
FMSIZE	Family size
IMR	Inverse Mill's ratio variable
LHIUSEDM	Same as LS2 (see below)
LS2	Dummy variable for Category 2 households (net buyers of labour)
LS2XPFNT	LS2 interacted with PFNLTR42
LS2xPFNY	LS2 interacted with PFNLYR42
LS2xRWGM	LS2 interacted with RLWGRTM2
LS3	Dummy variable for Category 3 households (autarchic)
LS3xPFNY	LS3 interacted with PFNLYR42
LS3xRWGM	LS3 interacted with RLWGRTM2
LS1	Dummy variable for Category 1 households (net sellers of labour)
NOC05	Number of children aged 0 to 5 in household
NOC69	Number of children aged 6 to 9 in household
PFNLYR42	Real per capita non-labour income, including net farm profit, (deducting value of own family labour at market wage rates)
PFNLTR42	Real per capita non-labour income, including net farm profit, (deducting value of own family labour at effective wage rates given by RLWGF2TH and RLWGM2TH)
PHASEDM	Dummy for the winter seasonal phase of survey data (2 phases)
RLWGF2TH	= RLWGRTF2 for LS1 category households
	= (RLWGRTF2 divided by θ) for LS2 category households
RLWGM2TH	= RLWGRTM2 for LS1 category households
	= (RLWGRTM2 divided by θ) for LS2 category households
RLWGRTF2	Real daily wage rate for female hired labour (kg. of paddy)
RLWGRTM2	Real daily wage rate for male hired labour (kg. of paddy)
UNMARRDM	Dummy for unmarried person
Dependent variable	= Total days of work on own farm or off farm, reported on a monthly basis, for each of two six-monthly seasonal cropping cycles.

Appendix Table 7A.2   Complete Regression Results: Male Labour Supply All Categories Model B2

Data subset:	Males, All Categories (1,2,3) Sample N = 2,542
Estimation method:	OLS (heteroskedasticity consistent errors)
Model description: (**B2**)	Common market wage;
	with intercept & slope dummies

Variable	estimated coefficient	standard error	t-ratio	p-value
RLWGRTM2	6.301	1.965	3.21	*0.001*
RLWGRTF2	-4.529	1.910	-2.37	*0.018*
FMSIZE	1.313	0.328	4.00	*0.000*
PFNLYR42 (x100)	-3.307	0.481	-6.88	*0.000*
FMNOMLF2	-9.782	0.949	-10.31	*0.000*
FMNOFLF2	0.488	0.921	0.53	*0.597*
AGE	1.052	0.375	2.81	*0.005*
AGESQ	-0.021	0.005	-4.53	*0.000*
DISDM21	9.00	1.698	5.30	*0.000*
DISDM23	-29.55	3.233	-9.14	*0.000*
PHASEDM	-4.16	1.271	-3.27	*0.001*
DMRLHH1	4.06	4.030	1.01	*0.314*
DMRLHH3	1.70	3.523	0.48	*0.630*
DMRLHH4	4.50	6.455	0.70	*0.486*
DMRLHH6	-3.06	3.935	-0.78	*0.436*
EDUCYR	-0.36	0.455	-0.79	*0.428*
UNMARRDM	-11.71	2.859	-4.08	*0.000*
ETHHLDM	4.47	2.483	1.80	*0.072*
ETHTLDM	4.87	1.949	2.50	*0.012*
CONS	110.5	9.561	11.56	*0.000*
Own wage slope dummies				
Catg. 2 LS2xRWGM	3.568	1.839	1.94	*0.052*
Catg. 3 LS3xRWGM	-5.702	1.651	-3.46	*0.001*
Income slope dummies (x100)				
Catg. 2 LS2xPFNY	2.938	0.5	5.88	*0.000*
Catg. 3 LS3xPFNY	3.746	0.65	5.77	*0.000*
Intercept dummies				
Catg. 2 LS2	-50.5	7.986	-6.34	*0.000*
Catg. 3 LS3	-22.9	8.206	-2.80	*0.005*

Adjusted R-Square	0.41
Standard error of the estimate (Sigma)	32.22
Breusch-Pagan Heteroskedasticity Test = 148.8 (χ^2 with 25 df)	
Ramsey RESET Specification Test (Reset2) = 6.179 (F with 1, 2515 df)	

See Appendix Table 7A.1 for variable definitions.

Appendix Table 7A.3 Complete Regression Results: Male Labour Supply Model C

Data subset:	Males Categories 1 & 2 (excludes autarchic) Sample N = 2,027
Estimation method:	OLS with White's heteroskedasticity consistent errors and error correction for estimated θ
Model description: (C)	Varying effective wage; no wage slope dummies

Version	1			2			
	with female wage			without female wage			
Variable	coeff.	stand. error	t-ratio	coeff.	stand. error	t-ratio	p-value
RLWGM2TH	5.687	1.614	3.52	4.521	1.417	3.19	0.001
RLWGF2TH	-1.691	1.604	-1.05	--	--	--	--
FMSIZE	1.044	0.378	2.76	1.066	0.462	2.31	0.021
PFNLTR42 x 100	-3.125	0.496	-6.30	-3.143	0.497	-6.32	0.000
FMNOMLF2	-9.433	1.099	-8.58	-9.476	1.136	-8.34	0.000
FMNOFLF2	1.896	1.014	1.87	1.870	1.037	1.80	0.072
AGE	1.398	0.421	3.32	1.391	0.421	3.31	0.001
AGESQ	-0.023	0.005	-4.59	-0.023	0.005	-4.57	0.000
DISDM21	8.325	1.792	4.65	8.029	1.894	4.24	0.000
DISDM23	-25.426	4.086	-6.22	-24.849	4.192	-5.93	0.000
PHASEDM	-4.334	1.393	-3.11	-4.335	1.394	-3.11	0.002
DMRLHH1	4.033	4.463	0.90	4.370	4.448	0.98	0.326
DMRLHH3	1.274	4.120	0.31	1.530	4.122	0.37	0.711
DMRLHH4	9.086	7.072	1.29	9.270	7.097	1.31	0.192
DMRLHH6	-4.774	4.526	-1.06	-4.274	4.491	-0.95	0.341
EDUCYR	0.065	0.500	0.13	0.079	0.502	0.16	0.875
UNMARRDM	-7.332	3.337	-2.20	-7.264	3.350	-2.17	0.030
ETHHLDM	3.065	3.119	0.98	3.104	3.118	1.00	0.320
ETHTLDM	4.425	2.110	2.10	4.351	2.143	2.03	0.042
CONS	94.1	9.810	9.59	92.400	10.630	8.69	0.000
Category 2 Income slope dummy							
LS2XPFNT x 100	2.676	0.511	5.24	2.696	0.514	5.25	0.000
Category 2 Intercept dummy							
LS2	-47.0	5.31	-8.85	-47.3	8.15	-5.80	0.000

Adjusted R Square	0.39	0.39
Standard error of the estimate (Sigma)	31.53	31.53
Breusch-Pagan Heteroskedasticity Test	44.03 (χ^2, df= 21)	35.03 (χ^2, df= 20)
RESET (2) Test (F statistic)	2.03 (df=1, 2004)	2.50 (df=1, 2005)
Model Selection Diagnostics		
Akaike Final Prediction Error	1005.7	1005.2
Schwartz Criteria	1071.8	1071.2

Note: Model C version 1 (with the female cross wage variable included as a regressor) corresponds to the summary results of Model C presented in Table 7.4. See Appendix Table 7A.1 for variable definitions.

Appendix Table 7A.4 Complete Regression Results:
Female Labour Supply Model C (Heckit)

Data subset:	Females, Categories 1 & 2 only (N = 1,827)
	(excluding autarchic and in-active persons)
Estimation method:	Heckit (with adjusted standard errors)
Model description: (C)	Varying effective wage; no wage slope dummy

Variable	estimated coefficient	heckit standard error	t-ratio	heckit + θ adj. standard error	t-ratio
RLWGF2TH	5.374	1.257	4.28	1.370	3.92
FMSIZE	2.885	1.160	2.49	1.160	2.49
PFNLTR42 (x 100)	-2.326	0.465	-5.00	0.466	-4.99
FMNOMLF2	-7.067	1.497	-4.72	1.499	-4.71
FMNOFLF2	-4.424	1.569	-2.82	1.569	-2.82
AGE	1.764	0.528	3.35	0.529	3.33
AGESQ	-0.031	0.007	-4.17	0.007	-4.15
DISDM21	15.107	2.655	5.69	2.660	5.68
DISDM23	-43.106	4.166	-10.35	4.195	-10.27
PHASEDM	-6.635	1.377	-4.82	1.377	-4.82
DMRLHH1	14.909	4.576	3.26	4.577	3.26
DMRLHH3	8.648	4.962	1.74	4.964	1.74
DMRLHH4	6.959	3.904	1.78	3.905	1.78
DMRLHH6	21.086	7.128	2.96	7.143	2.95
EDUCYR	-3.037	1.700	-1.79	1.708	-1.78
UNMARRDM	-3.293	4.953	-0.66	4.956	-0.66
LS2XPFNT (x 100)	1.812	0.059	3.49	0.521	3.48
ETHHLDM	7.691	3.215	2.39	3.215	2.39
ETHTLDM	3.411	2.561	1.33	2.564	1.33
NOC05	-3.695	1.311	-2.82	1.313	-2.82
NOC69	0.038	1.293	0.03	1.293	0.03
LHIUSEDM	-17.90	6.691	-2.68	6.782	-2.64
IMR	0.096	0.081	1.19	0.081	1.19
CONS	46.41	12.11	3.83	12.42	3.74

Adjusted R-Square	0.30
Standard error of the estimate (Sigma)	29.65
Log Likelihood	-8765.4
Breusch-Pagan Heteroskedasticity Test (χ^2, df = 23)	168.7
RESET (2) Test (F statistic with df = 1, 1802)	3.26
Model Selection Diagnostics	
Akaike Final Prediction Error	894.0
Schwartz Criteria	978.7

Note: The regression parameter estimates in this table correspond to the summary results of Model C (Heckit) presented in Table 7.9.2. See the Note to Table 7.9.2 for the description of the standard errors reported above. See Appendix Table 7A.1 for variable definitions.

Notes

1   The only direct connection between the production function estimation and the labour supply estimation is that the value of the θ parameter estimated in the former is used to create the effective wage and non-labour income variables for the labour supply regressions. Since the labour supply data module has not been utilized in generating the estimate of θ, it is an independent data set that can be checked for *conforming* evidence of θ being less than one.

2   The precise form of the wage gap created by the relative efficiency difference factor (θ) is illustrated in Figure 7.1 in Section 7.2. The labour supply model specification tests reported in this chapter cannot discriminate between alternative underlying explanations that give rise to Figure 7.1. But they can rule out other explanations that result in an arbitrary wage gap model inconsistent with Figure 7.1.

3   A fourth possible category is a household in which no family member works on own farm cultivation and all the labour demand is supplied by hired labour. This category will likely be a small group in most settings; and it has in effect been excluded from the selected sample for the farm household model estimation in this study because the production function estimation in Chapter 6 was restricted to cases where some family labour was supplied for own farm cultivation.

4   The theoretical inconsistency of both M and H being positive for a household ignores seasonal variation in labour supply and demand and the resulting changes in the net labour market exposure of a household over the annual cropping cycle.

5   As indicated in Equation 3. 20, in the general case θ* is a function of the optimum levels of F and H. In the case of the linear composite for effective labour, θ* becomes a constant, equal to the estimated value of θ (0.75) in Chapter 6, for all households. This value is independent of all other input levels as well, given the separability of the labour inputs in the primary production function as also indicated in Chapter 6.

6   There is a straightforward equivalence between the wage gap in Figure 7.1 and the price gap created by the difference between *c.i.f.* import and *f.o.b.* export prices in a model of international trade with transport costs. The small farmer who sells labour in the off farm market is exactly analogous to an exporting country where the domestic supply price is the *f.o.b.* price. The big farmer who "imports" hired labour is equivalent to an importing country that has a domestic supply price given by the higher *c.i.f.* price. In between is an autarchic country that does not trade if its domestic supply price is in between the *f.o.b.* and the *c.i.f.* price. (See Dixit and Norman, 1980).

7   This assumes no quantity constraints to the supply of hired labour at the market wage.

8   Bardhan and Rudra (1981) provide details of the terms of various forms of "attached" labour contracts in West Bengal. A similar variety of contractual modes also occur in the Nepal *tarai* region for which data was not collected in the MPBHS. Hence the labour supply of farm servants cannot be separately modelled in this study.

9   The specific occupation codes used to restrict the sample to agricultural workers, given the detailed occupation coding list of the MPHBS, were codes 61 (farmers) and 62 (agricultural and animal husbandry workers) and 99 (common labourers). The last category would include some non-agricultural labourers, such as construction workers and porters, but their wage rates are likely to be closely related to the daily wage rate for farm labourers.

10 This is a very small proportion of the final sample of almost 1,300 active individuals used in the male labour supply regressions.

11   The real wage rates reported in Table 7.2 are computed by deflating the nominal wage reported to be paid out to hired labour by the average price of paddy for each sample village. The mean values in Table 7.2 for the real wage rate (and other variables) differ slightly for the male and female worker samples since they are unweighted averages summed over working individuals, and not over the common set of sample households.

12   The sample of individuals included in the labour supply regressions is further restricted to family members aged 15 to 60 who report their main occupation as being agricultural workers or own account farm operators.

13   See Section 5.3.2 in Chapter 5. The variable PNLY in Equation 7.2 is NLY in Equation 5.3 divided by family size and the village specific price of paddy. This converts a nominal household level variable to a real per capita level variable.

14   Since rural labour markets are very localized, the wage reported to be paid out by a labour hiring household is likely to be the wage received by the neighbouring households that work on the hired labour market. In such a setting the fixed costs and other transaction costs involved in working on the neighbour's farm will tend to be negligible. See Section 5.3.2 in Chapter 5 for a fuller discussion of this issue.

15   In the *tarai* region of Nepal rice is the major food consumed. In many *tarai* areas wages of farm labourers are often paid in kind in units of paddy. The wages reported in the MPHBS are in part derived from the monetary conversion of wages paid in paddy.

16   For instance, consider a sample household with two economically active male workers and one female worker, where only one of the male individuals reports working on the off farm labour market. The market wage rate received by this male worker is also assigned as the effective wage rate for the other male worker. The hired labour market wage rate for female labour is assigned as the effective wage rate for the female member, even though she herself has not worked in the hired labour market.

17   If PNLY and all other variables in Models B and C were the same then the coefficient β^1 in Eq. 7.3B.1 would be the same as β in Eq. 7.3C.1, and β^2 in Eq. 7.3B.2 would be equal to β/θ in Eq. 7.5C.2. Hence Model B and Model C could not be distinguished.

18   The standard errors reported in Table 7.3 are OLS errors with White's heteroskedasticity consistent adjustments. The two step error correction is not required because all variables are based on the observed market wage rate.

19   The insignificant own wage coefficient for individuals in autarchic households is based on a specification where the cross wage effect for male labour supply is significantly negative, even for autarchic households. If the female wage rate variable is dropped from Model B2, the own wage effects of male labour supply are still significantly positive for Category 1 and 2 households. For autarchic households the own wage coefficient becomes negative, but its value is insignificant. So whether the female wage cross effect is included or not, the inference from Model B2 is clear. The labour supply behaviour of individuals in autarchic households is insensitive to the observed market wage rate, but a significant positive effect occurs for other household categories.

20   The critical value of the Wu-Hausman test statistic (which is χ^2 with 25 degrees of freedom) is 37.65 at the 5% level of significance, and 34.38 at the 10% significance level. Since the computed values of the test statistic in Table 7.3 are less than these critical values, the null of exogeneity is not rejected.

21   The autarchic households have 515 records for male workers in a total of 2,542 person-season records in the full sample. Dropping records from the autarchic households does not lead to specification bias of the estimates in Table 7.4 since the labour supply coefficients are assumed to be different for individuals in the autarchic households.

22 Some of the actual dummy variable coefficients are reported in Appendix 7 Tables where the full regression results are given for several model specifications.

23 In Model D the coefficient on the dummy variable for the own wage slope for Category 2 households is minus 0.129, with an adjusted standard error of 1.68. This value is not significantly different from zero.

24 The standard procedure of Davidson and MacKinnon's (1981) *J* test for two competing non-nested models is to run an extended regression for each model that includes as an extra regressor the predicted values from the competing model. Model selection diagnostic is based on the significance of the coefficient on the predicted value variables. This test does not always guarantee that one model will be preferred to the other (Maddala 1992: 515).

25 The Wu-Hausman test statistic for the *joint* exogeneity of the non-labour income and effective wage rates in Model C of Table 7.4 was 24.22. This is less than the critical value of χ^2 with 21 degrees of freedom at the 5% and 10% significance levels. The test statistics for the exogeneity of the non-labour income and the effective wage rates separately were also insignificant.

26 The standard errors in Table 7.5 are based on OLS with White's heteroskedasticity adjustments only. They do not account for θ being a pre-estimated parameter. The two step error correction has little bearing on the model selection diagnostics such as the *J* tests because the adjustment in the standard error of specific parameters are very minor. Small changes occur in the standard error for the real wage variable only.

27 Standard errors of the elasticity values noted in Table 7.6 for Model C are based on the two step adjusted covariance matrix, accounting for the fact that that θ is a pre-estimated parameter. For the details of this procedure see Section 4.4 in Chapter 4.

28 This result indicates that the source of the significant RESET Test in Table 7.3 is the sample of individuals in autarchic households. Their labour supply behaviour is apparently not correctly captured by a linear specification, even when allowing for specific intercept and slope dummies with respect to the market wage rate and non-labour income variables. There is a strong possibility that individuals in autarchic households could be constrained on their off farm labour supply behaviour.

29 See Chapter 5, Section 5.3 for the definitions for economically active persons adopted in the MPHBS, and for the categories of work identified.

30 The source of the sample selection bias is not due to the lack of randomness in the reduced sample of the economically active individuals only. Rather, the problem occurs due to the correlation between the error terms in the model that determines labour market participation and the model that determines the amount of labour supplied. If there is no correlation between these two error terms, the inverse Mill's ratio variable is insignificant, and the estimates based on OLS are consistent (Heckman, 1979).

31 Derivation of the marginal effects of the probit model follows Greene (1997: 884-86).

32 The procedure for deriving the Heckit corrected standard errors in the second step is given in the Shazam Users Reference Manual (White, 1993: 262-65).

33 When Model B of Table 7.4 is estimated on the full sample of all female workers from the three households categories (N = 2,288), with appropriate intercept and slope dummies, the estimated wage coefficient for individuals in autarchic households is 0.525, with a standard error of 0.686. This estimate is not significantly different from zero; but the coefficients for the other two household categories are significantly positive. Also, the RESET(2) Test statistic (F with 1, 2264 df) value of 18.98 indicates substantial mis-specification in Model B for female workers from all three household categories (as was the case in Model B for male workers).

34 The precise theoretical restriction based on utility maximization is that the *compensated* cross wage effects for male and female labour supply should be symmetric. (See the derivation in Kawaguchi, 1994). For this equality to hold, it is not necessary that both of the uncompensated cross wage effects be zero, as is imposed by the specification where the cross wage variable is dropped from the labour supply equation for each gender. However, given that in the specific results obtained the income effects (the coefficients on the non-labour income variable) for both male and female workers are approximately the same, the symmetry of the compensated wage effect depends crucially on the symmetry of the uncompensated effect. Hence, if one uncompensated cross wage effect is zero, the other must also be close to zero in order to fulfil the equality of the compensated cross wage effects.

35 The labour supply regression estimation was done in Shazam; and the procedure adopted for deriving the Heckit corrected standard errors follows the Shazam Users Reference Manual (White, 1993: 262-65).

36 In these adjustments to the standard errors, each particular effect is taken to be additive. The errors reported under the *hetcov + θ adj.* column for the OLS estimates in Tables 7.9 and 7.10 are White's heteroskedasticity consistent errors (computed under the HETCOV option in Shazam) to which is added the positive definite matrix which results from the adjustment for the fact that θ is pre-estimated. (See Equation 4.15 in Chapter 4). Similarly, the standard Heckit error corrections and the θ adjustment corrections are added together in the *heckit + θ adj.* column.

37 The Wu-Hausman test for the exogeneity of the wage and income variables is not significant in the regressions for the female workers sample either.

38 As in the case of the regressions for the male workers, all estimated versions of Models B, C and D for female workers also allow for an intercept dummy and slope dummy for the non-labour income variable for Category 2 households. Therefore, the basis for choosing among the alternative model specifications for the female labour supply regressions will also depend solely on the definition of the effective wage rate variable and inclusion of the wage slope dummy variables.

39 The estimated coefficient in Model D of the female wage slope dummy variable for Category 2 households is 1.82 with an standard error of 2.1 under OLS, and 1.23 with a standard error of 2.2 in the Heckit specification. The *p*-values associated with the test on these coefficients being significantly different from zero are 0.34 and 0.54, respectively. Therefore the null of a common wage coefficient for all individuals, but based on a model with effective wage rates, is not rejected by a wide margin.

40 The elasticities reported in Table 7.12 are conditional elasticities. They indicate only the effect on the labour supply of women who are already participating in the labour market. In the Heckit specification the effect of, for instance, a wage change has two dimensions: it increases the probability of participation among inactive women, and it increase the workdays of those who are already working (McDonald and Moffitt, 1980). The former effect is ignored on the calculations of the elasticities in Table 7.12.

41 The difference arises because the mean levels of the average days of work are quite different for individuals in Category 2 and Category 1 households. See Table 7.2.

42 These labour supply estimates in effect represent the consumer equilibrium of the farm household expressed implicitly in a demand system with a composite consumption good and leisure. Hausman (1981) has derived the direct and indirect utility functions underlying a linear labour supply function. Stern (1986) also discusses the utility implications of a linear labour supply function.

8 Summary and Conclusions

8.1 Overview

The main research question addressed in this book is whether it is important to distinguish between family and hired labour as production inputs in the traditional peasant agricultural production systems of the southern plain region of Nepal. The prime motivation behind this question was to test the validity of the conventional specification of the farm household model that treats family and hired labour as homogeneous inputs, allowing the production and consumption decisions of farm households to be modelled recursively. Another motivation was to explain the variation in the intensity of labour use across farm size so commonly observed in traditional agriculture.

Farm household models that reflect the integrated complex behavioural responses of a joint producer and consumer agent are important analytical tools for policy makers in developing countries. A recursive farm household model, in which the production and consumption/labour supply decisions need not be modelled jointly, offers considerable advantages in empirical implementation. Econometric estimation of non-recursive models is analytically cumbersome and the behavioural parameters of interest often involve non-linearities in the estimating equation even if the underlying model is linear in both the production and labour supply components.[1] The recursive feature, therefore, has important practical advantages and increases the popularity of farm household models for policy applications. Nevertheless, it is important to test that the empirical basis for the recursive structure is indeed well founded in a wide range of country settings. While there are several other grounds under which the recursive property of farm household models breaks down, the main concern is the completeness of rural labour markets and competitive wage determination, and whether hired and family labour inputs are indeed homogeneous inputs in farm production.

When family and hired labour must be treated as heterogeneous inputs the recursive structure of production and consumption choices usually breaks down. There will be a separate supply and demand equilibrium for family labour in which the household is no longer a price taker. The labour supply of the household is affected by its production decisions; and similarly, its factor demands are affected by the household's consumption/leisure preferences.

Methodology

The estimation methodology adopted in this study follows a two step procedure. In the first step, an aggregate farm production function is estimated with which the heterogeneity of family and hired labour is tested. Several alternative specifications that allow for imperfect substitution between family and hired labour and varying marginal products are estimated. The tests for heterogeneity of the two types of labour are carried out through standard tests of statistical significance of the parametric restrictions that lead to a model with homogeneous labour inputs. The estimated parameters of the preferred production function are also used to derive the relevant factor demand elasticities that define the production side of the behavioural response of the farm household.

In the second step, a structural labour supply model is specified that is consistent with the type of labour heterogeneity detected in the production function estimation. This typically means that some of the variables used in the labour supply regressions will be derived from parameters estimated in the production function. The main variable of interest derived in this way is the appropriate shadow wage rate that reflects the true cost of family labour at the equilibrium labour supply position of different types of farm households.

A key question is how this shadow wage rate is related to the observed market wage rates for family and hired labour. The answer varies for different households depending on whether they are net buyers or sellers on the hired labour market. The labour supply equations are then estimated with these appropriate shadow, or effective, wage rates and with other variables that may also depend on the computed effective wage rates. The parameter estimates from these regressions, based on the correctly specified variables, then describe the labour supply behaviour of the farm household in a manner consistent with utility maximization and with the specific type of labour heterogeneity modelled in the production function.

The income and wage elasticities derived from the labour supply regression functions complete the set of parameters of the farm household model. Accurate estimates of the parameters of the labour supply function are of interest in themselves. However, these labour supply regressions can also be used to verify whether the labour supply model specification, based on the assumption of labour heterogeneity, is superior to the conventional specification, based on the assumption of homogeneous labour. In the latter specification the observed market wage rate for hired labour would also be an appropriate measure of the opportunity cost of family labour. Hence, the comparative statistical performance of alternative labour supply models which specify a common wage (consistent with family and hired labour being homogeneous inputs) or varying effective wages (consistent with labour heterogeneity) can be used to independently corroborate the production function based result on whether family and hired labour are homogeneous inputs in farm production.

The two step estimation procedure adopted in this book is a modification of the strategy proposed by Jacoby (1993) where this procedure is used for estimating a fully non-recursive farm household model in which family and hired labour are treated as completely separate inputs. In this study embedded in the first step estimation of a farm production function is a test for the heterogeneity of family and hired labour. In the second step, the labour supply function is specified and estimated in a manner that is theoretically consistent with the nature of any heterogeneity indicated in the production function estimation.

Herein lies the methodological novelty of this book. On the one hand, while the conventional approach to estimating farm production functions is simply to aggregate family and hired labour into a homogeneous total labour input, there are many exceptions which estimate production functions by treating family and hired labour as completely separate inputs, implying but not formally testing the extent to which they are imperfect substitutes. On the other hand, in labour supply estimations the conventional approach has been to assume that the observed market wage rate is the appropriate opportunity cost of labour for all individuals, including those who are self employed on the family farm. But a relatively small number of studies have used alternative derivations of a shadow wage variable to model the on farm component of the labour supply behaviour of farm household members. However, the two parts of these non-conventional treatments have not been combined directly in a theoretically consistent manner – i.e. to relate the shadow wage rates of family labour to the observed market wage rates for hired labour based on the extent of the efficiency differences between hired and family labour detected in the production function estimates. That was the task carried out in this study.

Data Source

The data utilized for the empirical component of this book come from a large nationally representative household budget survey conducted by Nepal Rastra Bank. The data collected in this survey has been recognized to be of a very high quality and comprehensive in its treatment of farm income, inputs and outputs (World Bank, 1990). The actual household sample used in this study is a subset of the national survey data limited to about 1,000 rural households (among who about 700 have operational land holdings) from the southern plain (*tarai*) region of Nepal.

In the *tarai* region there is a greater inequality of land ownership and higher incidence of hired labour use than in the farms of the northern hill and mountain regions of Nepal. The mainly subsistence farm households in the northern regions engage in a whole range of other ancillary activities apart from crop production. It is difficult to correctly specify the inputs and outputs for these other activities. As a result, the question of heterogeneity between family and hired labour in Nepalese agriculture is better addressed in the context of the *tarai* region.

The production function estimates are based on the aggregate annual output of crops grown by the land operating households. The labour supply regressions are based on the seasonal work days reported by all economically active adults aged 15 to 60 in the full sample of households (including the landless) whose main occupation was reported as an agricultural worker or own farm operator. This led to slightly more than 2,000 person-season labour supply records for the male family workers and more than 1,800 person-season records for the female workers in the final sample used for the labour supply regressions and model selection tests.

8.2 Summary of Analytical Results

There are three separate requirements that must be fulfilled for two production inputs, X_1 and X_2, if they are to be homogeneous:

i. X_1 and X_2 be separable[2] from other inputs in the production function;
ii. the elasticity of substitution between X_1 and X_2 is very large;
iii. the ratio of the marginal products of X_1 and X_2 is equal to one (at all levels of applications of X_1 and X_2). In other words, a unit increase in the application of X_1 has the same effect on output, *ceteris paribus*, as a unit increase in X_2 has.

The estimation results with respect to family and hired labour reported in this book do not reject conditions *a* and *b;* but *c* is clearly rejected.

The tests for the heterogeneity between family and hired labour were carried out in Chapter 6, using a translog production function specification. Two different sets of estimations were carried out. The first set was restricted to a subset of the sample households that reported the use of *both* family and hired labour in crop production. For this sample subset the translog production function was estimated with family and hired labour as two distinct inputs in addition to three other inputs (land, bullock power, and material inputs) which were interacted with the two labour inputs, following the conventional translog functional specification.[3]

The results from the first set of estimations clearly indicated that family and hired labour inputs were weakly separable from the three other inputs. Stronger forms of separability, including the Cobb-Douglas restriction on the translog production function, were rejected. The weak separability result means that a unit change in the application of the latter three inputs leaves unchanged the ratio of the marginal products of family and hired labour. Another key finding was that the estimated value of the marginal product of family labour was statistically higher than the marginal product of hired labour, computed at the geometric mean of the data; and the marginal product of hired labour was close to the observed market wage rates.

The weak separability of the labour inputs in the production function implies that the two types of labour can be consistently aggregated into a composite labour input. In the second set of estimations, several alternative functional forms for the composite labour aggregator function were specified. A translog function with a nested aggregate labour input was estimated utilizing the entire sample of land cultivating households, including those that did not use any hired labour input.

The empirical results showed that the preferred labour input aggregator function was a linear composite given by $Le = F + \theta H$, where Le is effective or composite labour, F is family labour work days and H is hired labour days. The estimated value of θ was 0.75, and it was shown to be significantly less than one. This preferred form implies that, although family and hired labour are perfect substitutes in farm production, they are not equally productive. When both are measured conventionally in time units, the application of an extra unit of family labour has a larger effect on output than an extra unit of hired labour. When hired and family labour are measured in *effective* units, one unit of hired labour substitutes for 0.75 units of family labour.

The higher productivity of family labour inputs can be readily justified on the grounds that the *effort* applied per unit of time is likely to be lower for hired than for family labour when supervision of hired labour is costly (Feder, 1985). Another reason could be that family labour acquires some farm-specific experience, which leads to higher labour productivity (Rosenzweig and Wolpin, 1985). However, the estimation work of this study was not designed to discriminate between alternative explanations for the lower productivity of hired labour. The detailed information required to do so was not available in the survey data used.

A key implication of the Le = F + θH labour aggregator function is that the difference in productivity between hired and family labour is constant and unaffected by the levels of other inputs. The analytical structure of the farm household model with heterogeneous labour inputs derived in Chapter 3 showed that under such conditions the farm household model is still recursive. The only difference with the conventional model with homogeneous wages is that the effective wage rates for family labour will differ according to the labour market exposure of the household in the hired labour market.

For a landless (or a small farm) household, which at the margin supplies labour on the hired labour market, the effective wage rate that determines its total labour supply equilibrium will be the observed market wage rate, w, for hired labour. For big farm households that are net buyers of hired labour, the effective wage rate they face for the supply of their own family labour is w/θ, since at the margin one unit of family labour can substitute for $1/\theta$ units of hired labour.

The result that θ is less than one in the production function estimation implies it would be irrational for the sample households to hire in and hire out labour simultaneously. This would be inconsistent with the higher productivity of family labour. There would be efficiency gains from transferring the hired out labour into own farm cultivation, substituting for the hired in labour. This prediction is borne out by the sample data. Only about 3% of the approximately 700 land-operating households report hiring in labour as well as some member of the household working on the hired labour market. Given that the survey data refers to the entire annual cropping cycle, and given the very time specific nature of agricultural operations and the strict gender-related division of labour, this is a striking result which is consistent with the efficiency difference between family and hired labour.

The estimation results for the labour supply component of the farm household model, reflecting the θ efficiency difference between family and hired labour, was presented in Chapter 7. The labour supply variable includes all reported workdays on own farm cultivation as well as hired labour market work over each of two seasonal survey rounds. The labour supply regression equations were estimated separately for individual male and female family members without imposing any cross restrictions. A probit sample selection correction was made for the labour supply regression equation for female workers to take account of the large proportion of women who report to be economically inactive.

The labour supply regression results with the θ adjusted effective wage rates (Model C) are very reasonable for both the male and female regressions. The own wage elasticities of labour supply are positive (0.2 to 0.4 for male workers, and 0.3 to 0.6 for female workers). These values, though relatively high, are within the range of estimates obtained by others for farm households in developing countries. Labour supply decreases with the level of non-labour income (implying leisure is a normal good); but these elasticities are rather small (-0.07 to -0.1), which is again consistent with previous studies. The differences in the estimated elasticities between the two genders and between small (labour hiring out) and big farm (labour hiring in households) are also as expected. The absolute values of the wage and income elasticities for female workers are higher than for males. The uncompensated own wage elasticities are higher for big farmers who hire in labour than for small farmers and landless labourers. The effects of other variables on labour supply are also as expected. Workdays increase with age but at a decreasing rate. The number of young children reduces female workdays but has no effect on male labour supply. Family size and ethnic/caste group dummy variables are also significant.

Regarding tests of model selection, there are only slight differences in the overall fit of the various labour supply models specifications and in the values of the estimated parameters. But the varying effective wage model is clearly preferred over the common wage version on the basis of standard model selection diagnostics (J tests for non-nested models and the Akaike Information Criteria). This result holds for both the male and female labour supply regressions. These model selection tests offer an independent corroboration of the efficiency difference between family and hired labour inputs in the farm production process. The evidence that the labour supply behaviour conforms to the production function estimation results supports the conclusion that there is a genuine efficiency difference between family and hired labour. The labour supply results reduce the likelihood that the estimate of $\theta < 1$ in the production function is caused by other unobserved factors not taken into account in the production function estimations.

8.3 Some Implications

Apart from the methodological issue of the appropriate specification of a farm household model that allows for heterogeneity between family and hired labour, there are several other important implications of the finding that a substantial efficiency difference exists between family and hired labour as production inputs in the *tarai* region of Nepal.

An immediate implication is with regard to the measurement of farm level efficiency. A celebrated stylised fact about agricultural production in developing countries is that small farms are cultivated more intensively (i.e. with higher levels of variable inputs used per hectare) than bigger farms. This often leads to an observed inverse relationship between farm size and productivity in terms of output per hectare. The relatively greater application of inputs on smaller farms is most pronounced in the case of labour. Per hectare labour input on small farms is consistently higher than on bigger farms over a large range of farm sizes; and this result may occur even if yields on small farms are not higher (Berry and Cline, 1979).

The usual explanation for the higher labour input on smaller farms has been labour market imperfections which lead small farms to apply too much family labour to their own farms, relative to the prevailing market wage rates at which big farms hire in labour (Sen, 1975; Carter, 1992). Such "dualism" in traditional agriculture can occur if, for instance, wage rates are not bid down competitively to clear the hired labour market so that small farm family members are constrained in the amount of off farm work they can find. The labour allocation equilibrium on the small farm is then given by the equality of the marginal product of labour with the real opportunity cost of on farm work, at a level below the observed market wage rate. This type of market failure explains the higher labour intensity on small farms since their real cost of labour is lower than the market wage rate for hired labour applicable on the big farms.[4]

The efficiency difference between family and hired labour provides an alternative explanation for the relatively lower per hectare input of labour on the bigger farms, without relying on labour or other factor market failures. The efficiency difference means that the effective wage rate faced by big farmers who hire in labour is larger than the wage rate faced by small farmers who also work on the off farm labour market. The small farmer equates the marginal returns from farm cultivation to the market wage rate (w) at which he is able to work in the off farm labour market. The big farmer faces a higher effective wage rate (w/θ) than the small farmer does in terms of equivalent units of labour because of the lower productivity of hired labour. Although family labour is applied more intensively to substitute for the less efficient hired labour, there is a rising utility cost to increased family labour application. Consequently, total per hectare labour input, measured in conventional units, is higher on small farms cultivated solely by family members than on big farms that rely mostly on hired labour, even when wages are competitively determined and the hired labour market clears.

It is important to distinguish between the factor market imperfection and the effort related efficiency difference explanations because they imply different opportunity costs of labour in agriculture. For instance, under the factor market imperfection hypothesis, if a member of a small farm household migrates to an urban area the farm output loss will be minimal. The marginal product of family labour on the small farm is hypothesized to be less than the rural market wage rate. Also, given the presumed factor market imperfections, other family members who are constrained on the off farm labour market can possibly supply extra labour to make up for the migrating member. Under the efficiency difference hypothesis, marginal products and the real cost of leisure foregone are equated to the effective wage rates on both small and big farms, with the proviso that the effective wage rate is higher on the big farms. So the withdrawal of family labour from either the big or small farm will involve a higher opportunity cost.[5]

Theoretically, both factor market imperfections and efficiency differences could exist simultaneously. One does not rule out the other. The main focus of this book was not to discriminate between alternative explanations for dualism in traditional agriculture. Nevertheless, the estimation results of this study, which clearly indicate labour heterogeneity, also provide strong evidence against the factor market imperfection hypothesis. Based on the estimated production function parameters in Chapter 6, the marginal product of family labour on small farms is approximately equal to the market wage rate. The labour supply regression results in Chapter 7 showed the market wage accurately reflects the opportunity cost of family labour in small farms (Category 1 households, which also includes the landless). There is no direct evidence that shortage of work at the going off farm wage forces members of small farm households to devote all their labour supply to the family farm.

Another important implication of the labour heterogeneity finding which has significant policy relevance is that major aggregate outcomes in the agricultural sector, for instance, total production, total labour absorption, and equilibrium rural wage rates, would be sensitive to changes in the distribution of individual household endowments of land and labour.

For instance, if family labour is more productive because it applies more effort per unit time than hired labour, then a re-distributive land reform program which transferred land to small family-labour-operated farms from big farms relying primarily on hired labour would increase the average labour intensity of cultivation. It would increase the total absorption of labour in the agricultural sector and agricultural output without the necessity of drawing any additional resources into the agricultural sector.

8.4 Suggestions for Further Research

The main shortcoming of the research presented in this study is that the explicit source of the efficiency difference detected between family and hired labour has not been analyzed directly. From the available data set it is not possible to discriminate between alternative explanations for the higher efficiency of family labour – i.e. differences in the intensity of effort applied, farm specific-experience of older generations, or some other unobservable heterogeneity. Data sets with more detailed information on specific characteristics of family and hired labour employed on each farm and the specific tasks they do may be able to discriminate among these alternative explanations. A key variable may be the supervisory role that family labour plays over hired labour. The MPHBS data set from Nepal Rastra Bank did not distinguish between the actual physical fieldwork and supervisory role of family labour inputs. Where this distinction is available in the data, if it can be shown that the efficiency of hired labour is improved with greater supervisory input from family members, as in (Frisvold, 1994), then it would provide more direct evidence for the shirking-based explanation for the lower efficiency of hired labour detected in this study.

The potential importance of the supervisory role of family labour in the presence of hired labour inputs also suggests that family labour may provide a different type of labour service – one which combines elements of "management services" with units of ordinary labour. While the latter component could be easily substitutable with hired labour, the management services component is likely to be imperfectly substitutable. Further empirical research work with appropriate data sets that can distinguish between these two components of family labour would be a useful extension of this study. One viable way to deal with the problem of the essential jointness of the management services and ordinary labour units of family labour could be to estimate production functions with the labour input of the main family-based farm manager (or decision maker) as a separate labour input variable. This specification could then be used to test whether the farm manager's labour input is heterogeneous with respect to hired labour, or indeed with the labour input of other family members.

Another area for further research is to develop and estimate more elaborate specifications for the labour supply of individual household members which is consistent with labour heterogeneity. The labour supply regressions in Chapter 7 were based on a simple linear model. The prime interest was to find corroborating evidence for labour heterogeneity. Other interesting questions in the labour supply behaviour of farm households – intra-family labour allocation rules, female labour supply decisions being conditional on male labour supply, allowing for individuals to be constrained in their off farm workdays, household fixed effects, etc. – which may require a more flexible specification of labour supply to wage and income changes, were not addressed. These issues and other more elaborate analytical structures and estimation procedures for modelling the labour supply behaviour of farm households, together with labour heterogeneity, would be a fruitful area for further research. This would provide more robust results on whether the effective wage rates for family labour varies according to the hired labour market exposure of the farm households, in the presence of other possible sources of varying wage and income responses of labour supply.[6]

Other refinements to the simple labour supply specification of this study can be made by considering a joint family labour supply model for individual members of a household, where the workdays of a single person is affected by the workdays put in by other household members according to some age-sex grouping (as in Newman and Gertler, 1994). Another adjustment would be to test for and then incorporate constraints on the number of days for which off farm work is available, given the seasonal nature of farm operations, in estimating the labour supply functions. It would be worthwhile to then test for behavioural responses conforming to labour heterogeneity with this more general specification.

Another fruitful line of extension of this research would be to embed the structure and parameters of the farm household model estimated in this study into a larger simulation model of the agricultural economy of the *tarai* region of Nepal. The simulation model could then look specifically at the household-specific welfare effects and aggregate outcomes of alternative land reform policies that re-distributed land from big farms dependent on hired labour to family based cultivators. The general equilibrium effects, allowing for wage rates to change in response to changes in land distribution, are unclear and worthy of analysis. Of particular interest would be household-specific welfare effects. Even if the land transfers *increased* the aggregate labour absorption, there could still be a *reduction* in demand for hired labour that would lead to a reduction in the market wage rates for hired labour, and hence adversely affect landless households that were not beneficiaries of the land transfers. The direction and exact magnitude of such general equilibrium effects will depend on the precise values of the elasticities of labour demand and supply for the different classes of farm households. The research work reported in this book has made an initial contribution towards specifying the appropriate analytical framework and carrying out the relevant estimation work to obtain accurate estimates of these elasticities and other relevant parameters required for such a simulation model.

Notes

1   This potential problem has been highlighted by Jacoby (1993: 908 footnote 5).
2   For a production process utilizing *n* inputs, *separability* of inputs X_1 and X_2 from the other inputs implies that the marginal rates of substitution between X_1 and X_2 are independent of the levels of the other *n*-2 inputs. This implies that the ratio of the marginal products of X_1 and X_2 is invariant to the level of the other inputs. This notion of "separability" is, of course, a different concept than the "separability" or recursive property of farm household models.
3   Each of the family and hired labour input subtotal is an aggregation of male and female labour within each category. This aggregation is done using the ratio of the reported village level female wage rate to the male wage rate for hired labour to convert female labour days into equivalent male labour days. The mean value of this ratio was 0.85 in the sample data.
4   This is sometimes referred to as a greater "self-exploitation" of peasant labour on small family farms (Sen, 1975).
5   Unless, of course, the household's equilibrium labour allocation with heterogeneous labour also satisfies the "surplus labour" equilibrium in the sense defined by Sen (1966), with a flat schedule for the real cost of family labour. (The real cost of labour measures the marginal rate of indifferent substitution between consumption and leisure, as represented by the VV curve in Figure 4.1B). In an initial equilibrium on a flat section of the real cost of labour schedule, the withdrawal of some family members from the agricultural labour force induces other family members to work longer hours, since there is no increasing cost to higher levels of labour supply by an individual. Surplus labour exists in the sense that some workers can be transferred out of the farm sector without reducing labour input and farm output.
6   Stern (1986) provides a menu of alternative functional forms for labour supply functions and their underlying properties that would be a useful staring point. Only a limited number of specifications allow for a flexible response of labour supply to wage and non-labour income changes.

Bibliography

Abdulai, Awudu and Punya P. Regmi (2000), Estimating Labour Supply of Farm Households under Nonseparability: Empirical Evidence from Nepal, *Agricultural Economics*, 22(3): 309-20.

Abedin, Joynal and G. K. Bose (1988), Farm Size and Productivity Difference - A Decomposition Analysis, *Bangladesh Development Studies*, 16(3): 71-9.

Abowd, J. and O. Ashenfelter (1981), Anticipated Unemployment, Temporary Layoffs, and Compensating Wage Differentials, in S. Rosen (ed), *Studies in Labour Markets*, Chicago: Univ. of Chicago Press, 141-71.

Acharya, Meena (1979), Surplus Labour in Nepalese Agriculture, *The Journal of Development and Administrative Studies* (Katmandu, Nepal), 1(2): 222-43.

Acharya, Meena (1987), A Study of Rural Labour Markets in Nepal, unpublished Ph.D. dissertation, University of Wisconsin, Madison.

Adulavidhaya, K., Y. Kuroda, L. Lau, P. Lerttamrab and P. Yotopoulos (1979), A Microeconomic Analysis of the Agriculture of Thailand, *Food Research Institute Studies*, 17: 79-86.

Ahmed, Iqbal (1981), Farm Size and Labour Use: Some Alternative Explanations, *Oxford Bulletin of Economics and Statistics*, 43(1): 73-88.

Alderman, Harold and D. Sahn (1993), Substitution between Goods and Leisure in a Developing Country, *American Journal of Agricultural Economics*, 75(4): 875-83.

Arrow, K., H. Chenery, B. Minhas and R. Solow (1961), Capital Labour Substitution and Economic Efficiency, *The Review of Economics and Statistics*, 43(3): 225-47.

Bapna, S. L., H. P. Binswagner and J. Quizon (1984), Systems of Output Supply and Factor Demand Equations for Semi-Arid Tropical India, *Indian Journal of Agricultural Economics*, 39(2): 179-202.

Bardhan, Kalpana (1993), Work in South Asia: An Inter-regional Perspective, in S. Raju and D. Bagchi (eds), *Women and Work in*

South Asia: Regional Patterns and Perspectives, London and New York: Routledge, 39-73.

Bardhan, Pranab K. (1984), Peasant Labor Supply: A Statistical Analysis, in P. K. Bardhan, *Land, Labor and Rural Poverty: Essays in Development Economics*, New York: Columbia University Press.

Bardhan, Pranab K. (1979), Labor Supply Functions in a Poor Agrarian Economy, *American Economic Review*, 69(1): 73-83.

Bardhan, Pranab K. (1973), Size, Productivity and Returns to Scale: An Analysis of Farm Level Data in Indian Agriculture, *Journal of Political Economy*, 81(6): 1370-80.

Bardhan, Pranab K. and Ashok Rudra (1981), Terms and Conditions of Labour Contracts in Agriculture: Results of a Survey in West Bengal, 1979, *Oxford Bulletin of Economics and Statistics*, 43(1): 89-111.

Barnum, H. and L. Squire (1979), *A Model of an Agricultural Household*, The World Bank, Washington D.C.

Barro, R. J. (1977), Unanticipated Money Growth and Unemployment in the United States, *American Economic Review*, 67(2): 101-15.

Becker, G. (1965), A Theory of the Allocation of Time, *Economic Journal*, 75(3): 493-517.

Belbase, Krishna and R. Grabowski (1985), Technical Efficiency in Nepalese Agriculture, *Journal of Developing Areas*, 19(4): 515-25.

Belbase, Krishna, R. Grabowski and O. Sanchez (1985), The Marginal Productivity of Inputs and Agricultural Production in Nepal, *Pakistan Development Review*, 24(1): 51-60.

Benjamin, Dwayne (1995), Can Unobserved Land Quality Explain the Inverse Productivity Relationship?, *Journal of Development Economics*, 46(1): 51-84.

Benjamin, Dwayne (1992), Household Composition, Labor Markets, and Labor Demand: Testing for Separation in Agricultural Household Models, *Econometrica*, 60(2): 287-322.

Berndt, Ernst R. and L. R. Christensen (1974), Testing for the Existence of a Consistent Aggregate Index of Labor Inputs, *American Economic Review*, 64(3): 391-404.

Berndt, Ernst R. and L. R. Christensen (1973a), The Internal Structure of Functional Relationships: Separability, Substitution and Aggregation, *Review of Economic Studies*, 40(3): 403-10.

Berndt, Ernst R. and L. R. Christensen (1973b), The Translog Function and the Substitution of Equipment, Structures, and Labor in U.S. Manufacturing 1929-68, *Journal of Econometrics*, 1(1): 81-113.

Berndt, E. and B. C. Field (ed) (1981), *Modelling and Measuring Natural Resources Substitution*, Cambridge, Mass.: MIT Press.

Berry, R. A. and W. Cline (1979), *Agrarian Structure and Productivity in Developing Countries*, Baltimore: Johns Hopkins Univ. Press.

Bhalla, Surjit and P. Roy (1988), Mis-specification in Farm Productivity Analysis: The Role of Land Quality, *Oxford Economic Papers*, 40(1): 55-73.

Binswanger, H. P. and M. R. Rosenzweig (1986), Behavioural and Material Determinants of Production Relations in Agriculture, *Journal of Development Studies*, 22(3): 503-39.

Binswanger, H. P. and M. R. Rosenzweig (1984), *Contractual Arrangements, Employment and Wages in Rural Labour Markets in Asia*, New Haven: Yale Univ. Press.

Binswanger, H. P. (1974), A Cost Function Approach to the Measurement of Elasticities of Factor Demand and Elasticities of Substitution, *American Journal of Agricultural Economics*, 56(2): 377-86.

Blackorby, Charles and R. Russell (1989), Will the Real Elasticity of Substitution Please Stand Up? (A Comparison of the Allen/Uzawa and Morishima Elasticities), *American Economic Review*, 79(4): 883-88.

Blackorby, Charles, D. Primont and R. Russell (1979), *Duality, Separability, and Functional Structure: Theory and Economic Applications*, New York: North Holland.

Blomquist, N. Soren (1983), The Effect of Income Taxation on the Labor Supply of Married Men in Sweden, *Journal of Public Economics*, 22(2): 169-97.

Blundell, Richard and C. Meghir (1986), Selection Criteria for a Microeconometric Model of Labour Supply, *Journal of Applied Econometrics*, 1(1): 55-80.

Blundell, Richard and I. Walker (1982), Modelling the Joint Determination of Household Labour Supplies and Commodity Demands, *Economic Journal*, 92(366): 351-64.

Boisvert, R. (1982), The Translog Production Function: Its Properties, Its Several Interpretations and Estimation Problems, *Agricultural Economics Research*, No. 82-28, Cornell University, Ithaca, N.Y.

Burtless, Gary and J. Hausman (1978), The Effect of Taxation on Labor Supply: Evaluating the Gary Negative Income Tax Experiments, *Journal of Political Economy*, 86(6): 1103-30.

Caillavet, F., H. Guyomard and R. Lifran (eds) (1994), *Agricultural Household Modelling and Family Economics*, Amsterdam: Elsevier.

Carter, M. R. (1984), Identification of the Inverse Relationship between Farm Size and Productivity: An Empirical Analysis of Peasant Agricultural Production, *Oxford Economic Papers*, 36(1): 131-46.

Chambers, R. (1986), *Applied Production Analysis: A Dual Approach*, Cambridge: Cambridge University Press.

Chayanov, A. V. (1966), *The Theory of Peasant Economy* (translation edited by Daniel Thorner, Basile Kerblay and R. E. F. Smith), Homewood, Ill.: R. D. Irwin (Published for the American Economic Association).

Chiappori, P. A. (1997), Introducing Household Production in Collective Models of Labor Supply, *Journal of Political Economy*, 105(1): 191-209.

Chiappori, P. A. (1988), Rational Household Labor Supply, *Econometrica*, 56(1): 63-90.

Cogan, John F. (1981), Fixed Costs and Labor Supply, *Econometrica*, 49(4): 945-63.

Cooke, Priscilla A. (1998), Intrahousehold Labor Allocation Responses to Environmental Good Scarcity: A Case Study from the Hills of Nepal, *Economic Development and Cultural Change*, 46(4): 807-30.

Davidson, R. and J. G. MacKinnon. (1981), Several Tests for Model Specification in the Presence of Alternative Hypotheses, *Econometrica*, 49(3): 781-93.

Deaton, Angus (1997), *The Analysis of Household Surveys: A Microeconometric Approach to Development Policy*, Baltimore: Johns Hopkins Univ. Press (for the World Bank).

Deaton, Angus (1984), Issues in the Methodology of Multi-market Analysis of Agricultural Pricing Policies, Discussion Paper No. 116, Research Program in Development Studies, Princeton University.

Deaton, Angus and John Muellbauer (1980), *Economics and Consumer Behaviour*, Cambridge: Cambridge University Press.

Diamond, C. and T. Fayed (1998), Evidence on Substitutability of Adult and Child Labour, *Journal of Development Studies*, 34(3): 62-70.

Denny, M. and M. Fuss (1977), The Use of Approximation Analysis to Test for Separability and the Existence of Consistent Aggregates, *American Economic Review*, 67(3): 404-18.

Diewert, W. E. (1976), Exact and Superlative Index Numbers, *Journal of Econometrics*, 4(2): 115-45.

Diewert, W. E. (1971), An Application of the Shephard Duality Theorem: A Generalized Leontief Production Function, *Journal of Political Economy*, 79(3): 481-507.

Diewert, W. E. and T. J. Wales (1989), Flexible Functional Forms and Global Curvature Conditions, *Econometrica*, 55(1): 43-86.

Dixit, Avinash K. and Victor Norman (1980), *Theory of International Trade: A Dual, General Equilibrium Approach*, Welwyn, Herts.: James Nisbet (Cambridge Economic Handbooks).

de Janvry, Alain, M. Fafchamps and E. Sadoulet (1991), Peasant Household Behaviour with Missing Markets: Some Paradoxes Explained, *Economic Journal*, 101(409): 1400-17.

Deolalikar, Anil B. (1988), Nutrition and Labor Productivity in Agriculture: Estimates for Rural South India, *Review of Economics and Statistics*, 70(3): 406-13.

Deolalikar, Anil B. and Wim P. M. Vijverberg (1987), A Test of Heterogeneity of Family and Hired Labour in Asian Agriculture, *Oxford Bulletin of Economics and Statistics*, 49(3): 291-305.

Deolalikar, Anil and Wim P. M. Vijverberg (1983), The Heterogeneity of Family and Hired Labor in Agricultural Production: A Test Using District-Level Data from India, *Journal of Economic Development*, 8(2): 45-69.

Douglas, Evan J. (1989), The Simple Analytics of the Principal-Agent Incentive Contract, *Journal of Economic Education*, 20(1): 39-51.

Feder, Gershon (1985), The Relation between Farm Size and Farm Productivity: The Role of Family Labor, Supervision and Credit Constraints, *Journal of Development Economics*, 18(2-3): 297-313.

Field, E. B. (1988), Free and Slave Labor in the Antebellum South: Perfect Substitutes or Different Inputs?, *The Review of Economics and Statistics*, 70(4): 654-9.

Frisvold, George B. (1994), Does Supervision Matter? Some Hypothesis Tests Using Indian Farm Level Data, *Journal of Development Economics*, 43(2): 217-38.

Ghose, A. (1979), Farm Size and Land Productivity in Indian Agriculture: A Reappraisal, *Journal of Development Studies*, 16(1): 27-49.

Goldman, S. and H. Uzawa (1964), A Note on Separability in Demand Analysis, *Econometrica*, 32(3): 387-99.

Grabowski, Richard (1990), Curve Fitting versus Hypothesis Testing: A Comment, *Applied Economics*, 22(4): 427-29.

Grabowski, R. and K. Belbase (1986), An Analysis of Optimal Scale and Factor Intensity in Nepalese Agriculture: An Application of a Ray-Homothetic Production Function, *Applied Economics*, 18(1): 1051-63.

Greene, William H. (1997), *Econometric Analysis* (3rd edition), New Jersey: Prentice Hall.

Griliches, Z. (1984), Economic Data Issues, in Z. Griliches and M. Intrilligator (eds), *Handbook of Econometrics*, vol. III, Amsterdam: North-Holland, 1466-514.

Gronau R. (1977), Leisure, Home Production and Work - The Theory of the Allocation of Time Revisited, *Journal of Political Economy*, 85(6): 1099-123.

Hall, R. E. (1973), Wages, Income and Hours of Work in the U.S. Labor Force, in G. G. Cain and H.W. Watts (eds), *Income Maintenance and Labor Supply*, New York: Academic, 102-62.

Hamal, Krishna (1991), Crop Productivity Analysis: Nepal, 1961-1987, unpublished Ph.D. dissertation, University of Alberta (Canada).

Hamermesh, Daniel (1986), Labor Demand in the Long Run, in Orley Ashenfelter and Richard Layard (eds), *Handbook of Labor Economics*, vol. I, Amsterdam: North-Holland, 429-71.

Harris, John R. and Michael P. Todaro (1970), Migration, Unemployment and Development: A Two Sector Analysis, *American Economic Review*, 60(1): 126-42.

Hausman J. A. (1978), Specification Tests in Econometrics, *Econometrica*, 46(6): 1251-71.

Hausman J. A. (1981), Labour Supply, in H. J. Aaron and J. A. Pechman (eds), *How Taxes Affect Economic Behaviour*, Brookings Institute, Washington D.C., 27-75.

Heckman, J. J. and Thomas Macurdy (1986), Labor Econometrics, in Z. Griliches and M. D. Intrilligator (eds.) *Handbook of Econometrics,* vol. III, Amsterdam: North-Holland, 1917-77.

Heckman J. J. (1979), Sample Selection Bias as a Specification Error, *Econometrica,* 47(1): 153-61.

Heckman J. J. (1980), Sample Selection Bias as a Specification Error, in James P. Smith (ed), *Female Labor Supply: Theory and Estimation*, Princeton: Princeton Univ. Press, 206-48.

Hicks, J. R. (1970), Elasticity of Substitution Again: Substitutes and Complements, *Oxford Economic Papers*, 22(3): 289-96.

Hicks, J. R. (1964), *The Theory of Wages* (2nd edition), London: Macmillan.

Hirashima, S. and M. Muqtada (1986), *Hired Labour and Rural Labour Markets in Asia*, ILO, Asian Employment Programme, New Delhi.

HMG/N, CBS (1997), *Statistical Yearbook Nepal*; His Majesty's Government of Nepal, Central Bureau of Statistics, Kathmandu.

HMG/N, CBS (1993), *Nepal Agricultural Census 1991/92 Report*, His Majesty's Government of Nepal, Central Bureau of Statistics, Kathmandu (various volumes).

HMG/N, NPC (National Planning Commission) (1983), *A Survey of Employment, Income Distribution and Consumption Patterns in Nepal*, His Majesty's Government of Nepal, National Planning Commission Secretariat, Kathmandu.

Hoque, Asarul (1988), Farm Size and Economic-Allocative Efficiency, *Applied Economics*, 20(10): 1353-68.

Huffman, Wallace (1980), Farm and Off Farm Work Decisions: The Role of Human Capital, *Review of Economics and Statistics*, 62(1): 14-23.

Jacoby, H. (1993), Shadow Wages and Peasant Family Labour Supply: An Econometric Application to the Peruvian Sierra, *Review of Economic Studies*, 60(4): 903-21.

Jacoby, H. (1991), Productivity of Men and Women and the Sexual Division of Labor in Peasant Agriculture of the Peruvian Sierra, *Journal of Development Economics*, 37(1-2): 265-87.

Jorgenson, D. and B. Fraumeni (1981), Relative Prices and Technical Change, in E. Berndt and B. C. Field (ed), *Modelling and Measuring Natural Resources Substitution*, Cambridge, Mass.: MIT Press, 17-47.

Kawaguchi, A. (1994), Testing Neoclassical and Non-neoclassical Models of Household Labour Supply, *Applied Economics*, 26(1): 9-19.

Killingsworth, M. (1983), *Labour Supply*, Cambridge: Cambridge Univ. Press.

Killingsworth, M. and J. Heckman (1986), Female Labor Supply: A Survey, in Orley Ashenfelter and Richard Layard (eds), *Handbook of Labor Economics*, vol. I, Amsterdam: North-Holland, 103-204.

Kuroda, Y. and P. Yotopoulos (1980), A Study of Consumption Behaviour of the Farm-Household in Japan: Application of Linear Logarithmic Expenditure System, *The Economic Review* (Japan), 31: 1-15.

Kuroda, Y. and P. Yotopoulos (1978), A Micro-economic Analysis of Production Behaviour of the Farm Household in Japan: A Profit Function Approach, *The Economic Review* (Japan), 29: 116-29.

Lambert, Sylvie and T. Magnac (1994), Measurement of Implicit Prices of Family Labour in Agriculture: An Application to Cote D'Ivoire, in F. Caillavet, H. Guyomard and R. Lifran (eds), *Agricultural Household Modelling and Family Economics*, Amsterdam: Elsevier, 9-24.

Lau, L., W. Lin, and P. Yotopoulos (1979), Applications of the Profit Function: Efficiency and Technical Change in Taiwan's Agriculture, *Food Research Institute Studies*, 17(1): 23-51.

Lau, L., W. Lin, and P. Yotopoulos (1978), The Linear Logarithmic Expenditure System: An Application to Consumption-Leisure Choice, *Econometrica*, 46(4): 843-68.

Laufer, Leslie A. (1985), The Substitution Between Male and Female Labor in Rural Indian Agricultural Production, *Yale Economic Growth Center Discussion Papers No. 472*, Yale University.

Leiderman, L. (1980), Macro-econometric Testing of Rational Expectations and Structural Neutrality Hypotheses for the United States, *Journal of Monetary Economics*, 6(1): 69-82.

Lopez, Ramon (1986), Structural Models of the Farm Household That Allow for Interdependent Utility and Profit-Maximization Decisions, in I. J. Singh, L. Squire and J. Strauss (eds), *Agricultural Household Models: Extensions, Applications, and Policy*, Baltimore: Johns Hopkins Univ. Press, 306-25.

Lopez, Ramon (1984), Estimating Labor Supply and Production Decisions of Self-Employed Farm Producers, *European Economic Review*, 24(1): 61-82.

Maddala, G. S. (1992), *Introduction to Econometrics* (2nd edition), New York: Macmillan.

McDonald, John and Robert Moffitt (1980), The Uses of Tobit Analysis, *The Review of Economics and Statistics*, 62(2): 318-21.

McFadden, Daniel (1978), Estimation Techniques for the Elasticity of Substitution and Other Production Parameters, in M. Fuss and D. McFadden (eds), *Production Economics: A Dual Approach to Theory and Applications*, vol. 2, Amsterdam: North Holland, 74-123.

Mahmood, Moazam and Nadeem ul Haque (1981), Farm Size and Productivity Revisited, *Pakistan Development Review*, 20(2): 151-90.

Mazumdar, Dipak (1989), Microeconomic Issues of Labor Markets in Developing Countries: Analysis and Policy Implications, EDI Seminar Paper No. 40., Economic Development Institute, World Bank, Washington D.C.

Mazumdar, Dipak (1975), The Theory of Share-Cropping with Labour Market Dualism, *Economica*, 42(167): 261-71.

Moffitt, Robert (1986), The Econometrics of Piecewise-Linear Budget Constraints: A Survey and Exposition of the Maximum Likelihood Method, *Journal of Business and Economic Statistics*, 4(3): 317-28.

Moghadam, Fatemet Etemad (1982), Farm Size, Management and Productivity: A Study of Four Iranian Villages, *Oxford Bulletin of Economics and Statistics*, 44(4): 357-79.

Moll, Peter G. (1990), Curve Fitting vs. Hypothesis Testing with Production Functions: Comment, *Applied Economics*, 22(1): 1-4.

Moschini, Giancarlo, D. Moro and R. Green (1994), Maintaining and Testing Separability in Demand Systems, *American Journal of Agricultural Economics*, 76(1): 61-73.

Mudhbary, P. (1988), A Demand System Analysis of Food Consumption in Nepal, unpublished Ph.D. dissertation, Michigan State University.

Murphy, Kevin and R. H. Topel (1985), Estimation and Inference in Two-step Econometric Models, *Journal of Business and Economic Statistics*, 3(4): 370-79.

Mundlak, Yair (1996), Production Function Estimation: Reviving the Primal, *Econometrica*, 60(2): 431-8.

Mundlak, Yair, (1968), Elasticities of Substitution and the Theory of Derived Demand, *Review of Economic Studies*, 35: 225-36.

Nakajima, Chihiro (1949), Internal Equilibrium Theory of the Farm Firm (in Japanese), *Journal of Rural Economics* (Japan), 21(3).

Nakajima, Chihiro (1986), *Subjective Equilibrium Theory of the Farm Household* (translated by R. Kada), Amsterdam: Elsevier.

Newman, John L. and Paul J. Gertler (1994), Family Productivity, Labor Supply, and Welfare in a Low Income Country, *Journal of Human Resources*, 29(4): 989-1026.

Newey, W. K. (1984), A Method of Moments Interpretation of Sequential Estimators, *Economics Letters*, 14: 201-6.

NRB (Nepal Rastra Bank) (1988), *Multipurpose Household Budget Survey A Study on Income Distribution, Employment and Consumption Patterns in Nepal*, Nepal Rastra Bank, Kathmandu.

NRB/ADB (1994), *Nepal Rural Credit Review Final Report* (in four volumes), Nepal Rastra Bank, Kathmandu, and the Asian Development Bank, Manila.

Pagan, Adrian (1986), Two Stage and Related Estimators and their Applications, *Review of Economic Studies*, 53(4): 517-38.

Pagan, Adrian (1984), Econometric Issues in the Analysis of Regressions with Generated Regressors, *International Economic Review*, 25(1): 221-47.

Pencaval, J. (1986), Male Labor Supply: A Survey, in Orley Ashenfelter and Richard Layard (eds), *Handbook of Labor Economics*, vol. I, Amsterdam: North-Holland, 3-102.

Pitt, Mark M. and M. R. Rosenzweig (1986), Agricultural Prices, Food Consumption, and the Health and Productivity of Indonesian Farmers, in I. J. Singh, L. Squire and J. Strauss (eds), *Agricultural Household*

Models: Extensions, Applications, and Policy, Baltimore: Johns Hopkins University Press, 153-82.

Pudney, S. (1989), *Modelling Individual Choice: The Econometrics of Corners, Kinks and Holes*, Oxford: Blackwell.

Quizon, J. and H. P. Binswagner (1986), Modeling the Impact of Agricultural Growth and Government Policy on Income Distribution in India, *The World Bank Economic Review*, 1(1): 103-48.

Quizon, J. and H. P. Binswanger (1983), Income Distribution in Agriculture: A Unified Approach, *American Journal of Agricultural Economics*, 65(3): 526-38.

Rauniyar, Krishna (1985), Labour Utilization and Non Farm Labor Supply Among Rural Farm Households: A Case Study of Hill and Tarai Districts, Research Paper Series No. 30, HMG/USAID/GTZ/Winrock Project, Kathmandu, Nepal.

Revankar, Nagesh (1971), A Class of Variable Elasticity of Substitution Production Functions, *Econometrica*, 39(1): 61-71.

Rosenzweig, M. R. (1988), Labor Markets in Low-Income Countries, in H. B. Chenery, T. N. Srinivasan and Jere R. Behrman (eds), *Handbook of Development Economics*, vol. I, Amsterdam: North Holland, 714-59.

Rosenzweig, M. R. (1980), Neoclassical Theory and the Optimizing Peasant: An Econometric Analysis of Family Labour Supply in a Developing Country, *Quarterly Journal of Economics*, 94(1): 31-55.

Rosenzweig, M. R. (1978), Rural Wages, Labor Supply, and Land Reform: A Theoretical and Empirical Analysis, *American Economic Review*, 68(5): 847-61.

Rosenzweig, M. R. and K. Wolpin (1985), Specific Experience, Household Structure, and Intergenerational Transfers: Farm Family Land and Labor Arrangements in Developing Countries, *Quarterly Journal of Economics*, 100(5), Supplement: 961-87.

Rudra, Ashok (1992), *Political Economy of Indian Agriculture*, Calcutta: Bagchi Publications.

Ryan, David L. and T. J. Wales (1998), A Simple Method for Imposing Local Curvature in Some Flexible Consumer Demand Systems, *Journal of Business and Economic Statistics*, 16(3): 331-8.

Sato, R. and T. Koizumi (1973), On the Elasticities of Substitution and Complimentarity, *Oxford Economic Papers*, 25(1): 44-56.

Seidman, L. (1989), Compliments and Substitutes: The Importance of Minding p's and q's, *Southern Economic Journal*, 56(1): 183-90.

Sen, Amartya K. (1975), *Employment, Technology and Development*, Oxford: Clarendon Press.

Sen, Amartya K. (1966), Peasants and Dualism with or without Surplus Labor, *Journal of Political Economy*, 74(5): 425-50.

Sharma, Kailash (1988), Farm-Household Economics and Some Empirics from the Semi-arid Tropics of India (unpublished manuscript), University of New England, Armidale, 1-11.

Sicular, Terry (1986), Using a Farm Household Model to Analyze Labor Allocation on a Chinese Collective Farm, in I. J. Singh, L. Squire and J. Strauss (eds), *Agricultural Household Models: Extensions, Applications, and Policy*, Baltimore: Johns Hopkins Univ., 277-305.

Sidhu, S. and C. Baanante (1981), Estimating Farm Input Demand and Wheat Supply in the Indian Punjab using a Translog Profit Function, *American Journal of Agricultural Economics*, 63(2): 237-46.

Singh, I . J., L. Squire and J. Strauss (eds) (1986), *Agricultural Household Models: Extensions, Applications, and Policy*, Baltimore: Johns Hopkins Univ. Press.

Singh, I. J., L. Squire and J. Strauss (1986c1), An Overview of Agricultural Household Models, Chapter 1 in I. J. Singh, L. Squire and J. Strauss (eds), *Agricultural Household Models: Extensions, Applications, and Policy*, Baltimore: Johns Hopkins Univ. Press, 17-47.

Singh, I. J., L. Squire and J. Strauss (1986c2), Methodological Issues, Chapter 2 in I. J. Singh, L. Squire and J. Strauss (eds), *Agricultural Household Models: Extensions, Applications, and Policy*, Baltimore: Johns Hopkins Univ. Press, 48-94.

Singh, I. J., L. Squire and J. Strauss (1986b), A Survey of Agricultural Household Models: Recent Findings and Policy Implications, *The World Bank Economic Review*, 1(1): 149-79.

Skoufias, Emmanuel (1994), Using Shadow Wages to Estimate Labor Supply of Agricultural Households, *American Journal of Agricultural Economics*, 76(2): 215-27.

Stern, Nicholas (1986), On the Specification of Labour Supply Functions, in R. Blundell and I. Walker (eds), *Unemployment, Search and Labour Supply*, Cambridge: Cambridge University Press, 143-89.

Strauss, John and Duncan Thomas (1995), Human Resources: Empirical Modeling of Household and Family Decisions, in T. N. Srinivasan and Jere R. Behrman (eds), *Handbook of Development Economics*, vol. IIIA, North Holland: Amsterdam, 1833-2023.

Strauss, J. (1986), The Theory and Comparative Statics of Agricultural Household Models: A General Approach (Appendix to Chapter 2) in I. J. Singh, L. Squire and J. Strauss (eds), *Agricultural Household*

Models: Extensions, Applications, and Policy, Baltimore: Johns Hopkins Univ. Press, 71-91.

Squires, Dale and S. Tabor (1994), The Absorption of Labor in Indonesian Agriculture, *Developing Economies*, 32(2): 167-87.

Tanaka, O. (1951), An Analyses of Economic Behaviour of the Farm-Household (in Japanese), *Journal of Rural Economics* (Japan), 22(4).

Topel, R. (1982), Inventories, Layoffs, and the Short Run Demand for Labor, *American Economic Review*, 72(4): 769-87.

Udry, Christopher (1996), Gender, Agricultural Production, and the Theory of the Household, *Journal of Political Economy*, 104(5): 1010-46.

Udry, Christopher, J. Hoddinott, H. Alderman and L. Haddad (1995), Gender Differential in Farm Productivity: Implications for Household Efficiency and Agricultural Policy, *Food Policy*, 20(5): 407-23.

Upadhyaya, Hari K. and Ganesh B. Thapa (1994), Modern Variety Adoption, Wage Differential, and Income Distribution in Nepal, in Cristina C. David and Keijiro Otsuka (eds), *Modern Rice Technology and Income Distribution in Asia*, Boulder, Colorado: L. Rienner Publishers, 281-323.

Uzawa, Hirofumi (1962), Production Functions with Constant Elasticities of Substitution, *Review of Economic Studies*, 29: 291-9.

Wallace, Michael B. (1989), Nepal: Food Pricing with an Open Border, in Terry Sicular (ed), *Food Price Policy in Asia*, Ithaca: Cornell Univ. Press, 183-243.

White, B. (1994), Children, Work and Child Labour: Changing Responses to the Employment of Children, *Development and Change*, 25(4): 849-78.

White, Kenneth J. (1993), *SHAZAM User's Reference Manual Version 7.0*, Toronto: McGraw Hill.

World Bank (1990), Nepal: Relieving Poverty in a Resource–Scarce Economy (in 2 volumes), World Bank Country Report No. 8635-NEP, Washington D.C.

World Bank (1989), *Nepal: Policies for Improving Growth and Alleviating Poverty*, World Bank Country Study, Washington D.C.

Wu, D. M. (1983), Tests of Causality, Pre-determinedness and Exogeneity, *International Economic Review*, 24(3): 547-58.

Zellner, A., J. Kmenta and J. Dreze (1966), Specification and Estimation of Cobb-Douglas Production Functions, *Econometrica*, 34(4): 789-95.

Index